# Aircraft Engineering: Design, Structures and Control Systems

# Aircraft Engineering: Design, Structures and Control Systems

Casey Stokes

Larsen & Keller
www.larsen-keller.com

Aircraft Engineering: Design, Structures and Control Systems
Casey Stokes
ISBN: 978-1-64172-139-4 (Hardback)



Larsen & Keller
Published by Larsen and Keller Education,
5 Penn Plaza,
19th Floor,
New York, NY 10001, USA

**Cataloging-in-Publication Data**

Aircraft engineering : design, structures and control systems / Casey Stokes.
p. cm.
Includes bibliographical references and index.
ISBN 978-1-64172-139-4
1. Aeronautics. 2. Airplanes--Design and construction. 3. Airframes.
4. Airplanes--Control systems. I. Stokes, Casey.
TL671.2 .A57 2019
629.13--dc23



Printed and bound in China.

For more information regarding Larsen and Keller Education and its products, please visit the publisher's website www.larsen-keller.com

# Table of Contents

# Preface

Aircraft design process is an engineering process concerned with the design of aircrafts according to standards of customer safety and comfort, technical specifications, and physical and economic constraints. Aircraft are designed according to specific purpose. Amphibious airplanes are designed so that they can operate both on land and water, while helicopters are engineered to be able to hover over an area over a period of time. Some other factors influencing it are space limitations, runway and safety issues, and pavement design. An understanding of aerodynamics, propulsion, controls, structure and mass is vital to aircraft design. Design is structured into the three phases of conceptual design, preliminary design phase and detail design phase. This book is a valuable compilation of topics, ranging from the basic to the most complex theories and principles in the field of aircraft design. It elucidates the concepts and innovative models around prospective developments with respect to aircraft design. This textbook, with its detailed analyses and data, will prove immensely beneficial to professionals and students involved in this area at various levels.

Given below is the chapter wise description of the book:

Chapter 1- An aircraft is a machine that is enabled to fly in the air, by using either static or dynamic lift. Aircraft can be of different types on the basis of their lift type, propulsion or usage. This is an introductory chapter, which will introduce briefly different varieties of aircraft such as airliner, powered aircraft, military aircraft, supersonic aircraft and experimental aircraft.

Chapter 2- Aerodynamics is the study of the motion of air around a solid object. This chapter has been carefully written to provide an easy understanding of the fundamentals of aerodynamics by elucidating topics such as compressible flow, center of pressure, boundary layer, supersonic speed and hypersonic speed, besides others.

Chapter 3- A propulsion system typically consists of a power source like an engine or motor, propellers, wheels and axles, and a propulsive nozzle for generating the force. This chapter discusses the principles of jet propulsion and also elucidates the different innovative designs of aircraft engine and jet engine.

Chapter 4- An aircraft flight control system consists of flight control surfaces, cockpit controls and the necessary operating mechanisms to control the direction of the flight. The topics elaborated in this chapter will help in presenting a better perspective of flight control surfaces and aircraft engine controls.

Chapter 5- Science and technology have undergone rapid developments in the past decade, which has resulted in innovation in aircraft parts, design and structures. Many significant topics in the field of aircraft designing such as airframe, fuselage, wing, propeller, landing gear, etc. have been extensively detailed in this chapter.

At the end, I would like to thank all those who dedicated their time and efforts for the successful completion of this book. I also wish to convey my gratitude towards my friends and family who supported me at every step.

Casey Stokes

1

# Aircraft and its Types

An aircraft is a machine that is enabled to fly in the air, by using either static or dynamic lift. Aircraft can be of different types on the basis of their lift type, propulsion or usage. This is an introductory chapter, which will introduce briefly different varieties of aircraft such as airliner, powered aircraft, military aircraft, supersonic aircraft and experimental aircraft.

## Aircraft

An aircraft is any machine supported for flight in the air by buoyancy or by the dynamic action of air on its surfaces, especially powered airplanes, gliders, and helicopters.

An aircraft is any machine capable of atmospheric flight.

A first division by design among aircraft is between lighter-than-air, aerostat, and heavier-than-air aircraft, aerodyne.

Examples of lighter-than-air aircraft include non-steerable balloons, such as hot air balloons and gas balloons, and steerable airships (sometimes called dirigible balloons) such as blimps (that have non-rigid construction) and rigid airships that have an internal frame.

In heavier-than-air aircraft, there are two ways to produce lift: aerodynamic lift and engine lift.

In the case of aerodynamic lift, the aircraft is kept in the air by wings or rotors.

With engine lift, the aircraft defeats gravity by use of vertical thrust.

Examples of engine lift aircraft are rockets, and VTOL aircraft such as the Hawker-Siddeley Harrier.

NASA test aircraft

Among aerodynamically lifted aircraft, most fall in the category of fixed-wing aircraft, where horizontal airfoils produce lift, by profiting from airflow patterns determined by Bernoulli's equation and, to some extent, the Coanda effect.

The Mil Mi-8 is the most-produced helicopter in history

The Cessna 172 Skyhawk is the most produced aircraft in history

*Voodoo*, a modified P 51 Mustang is the 2014 Reno Air Race Champion

## Methods of Lift

## Lighter than Air – Aerostats

Hot air balloons

Aerostats use buoyancy to float in the air in much the same way that ships float on the water. They are characterized by one or more large gasbags or canopies, filled with a relatively low-density gas such as helium, hydrogen, or hot air, which is less dense than the surrounding air. When the weight of this is added to the weight of the aircraft structure, it adds up to the same weight as the air that the craft displaces.

Small hot-air balloons called sky lanterns were first invented in ancient China prior to the 3rd century BC and used primarily in cultural celebrations, and were only the second type of aircraft to fly, the first being kites which were first invented in ancient China over two thousand years ago.

A balloon was originally any aerostat, while the term airship was used for large, powered aircraft designs – usually fixed-wing. In 1919 Frederick Handley Page was reported as referring to "ships of the air," with smaller passenger types as "Air yachts." In the 1930s, large intercontinental flying boats were also sometimes referred to as "ships of the air" or "flying-ships". – though none had yet been built. The advent of powered balloons, called dirigible balloons, and later of rigid hulls allowing a great increase in size, began to change the way these words were used. Huge powered aerostats, characterized by a rigid outer framework and separate aerodynamic skin surrounding the gas bags, were produced, the Zeppelins being the largest and most famous. There were still no fixed-wing aircraft or non-rigid balloons large enough to be called airships, so "airship" came to be synonymous with these aircraft. Then several accidents, such as the Hindenburg disaster in 1937, led to the demise of these airships. Nowadays a "balloon" is an unpowered aerostat and an "airship" is a powered one.

Airship USS Akron over Manhattan

A powered, steerable aerostat is called a *dirigible*. Sometimes this term is applied only to non-rigid balloons, and sometimes *dirigible balloon* is regarded as the definition of an airship (which may then be rigid or non-rigid). Non-rigid dirigibles are characterized by a moderately aerodynamic gasbag with stabilizing fins at the back. These soon became known as *blimps*. During the Second World War, this shape was widely adopted for tethered balloons; in windy weather, this both reduces the strain on the tether and stabilizes the balloon. The nickname *blimp* was adopted along with the shape. In modern times, any small dirigible or airship is called a blimp, though a blimp may be unpowered as well as powered.

## Heavier-than-Air – Aerodynes

Heavier-than-air aircraft, such as airplanes, must find some way to push air or gas downwards, so that a reaction occurs (by Newton's laws of motion) to push the aircraft upwards. This dynamic movement through the air is the origin of the term *aerodyne*. There are two ways to produce dynamic upthrust: aerodynamic lift, and powered lift in the form of engine thrust.

Aerodynamic lift involving wings is the most common, with fixed-wing aircraft being kept in the air by the forward movement of wings, and rotorcraft by spinning wing-shaped rotors sometimes called rotary wings. A wing is a flat, horizontal surface, usually shaped in cross-section as an aerofoil. To fly, air must flow over the wing and generate lift. A *flexible wing* is a wing made of fabric or thin sheet material, often stretched over a rigid frame. A *kite* is tethered to the ground and relies on the speed of the wind over its wings, which may be flexible or rigid, fixed, or rotary.

With powered lift, the aircraft directs its engine thrust vertically downward. V/STOL aircraft, such as the Harrier Jump Jet and F-35B take off and land vertically using powered lift and transfer to aerodynamic lift in steady flight.

A pure rocket is not usually regarded as an aerodyne, because it does not depend on the air for its lift (and can even fly into space); however, many aerodynamic lift vehicles have been powered or assisted by rocket motors. Rocket-powered missiles that obtain aerodynamic lift at very high speed due to airflow over their bodies are a marginal case.

## Fixed-wing

An Airbus A380, the world's largest passenger airliner

The forerunner of the fixed-wing aircraft is the kite. Whereas a fixed-wing aircraft relies on its forward speed to create airflow over the wings, a kite is tethered to the ground and relies on the wind blowing over its wings to provide lift. Kites were the first kind of aircraft to fly, and were invented in China around 500 BC. Much aerodynamic research was done with kites before test aircraft, wind tunnels, and computer modelling programs became available.

The first heavier-than-air craft capable of controlled free-flight were gliders. A glider designed by George Cayley carried out the first true manned, controlled flight in 1853.

Practical, powered, fixed-wing aircraft (the aeroplane or airplane) were invented by Wilbur and Orville Wright. Besides the method of propulsion, fixed-wing aircraft are in general characterized by their wing configuration. The most important wing characteristics are:

- Number of wings – Monoplane, biplane, etc.
- Wing support – Braced or cantilever, rigid, or flexible.
- Wing planform – including aspect ratio, angle of sweep, and any variations along the span (including the important class of delta wings).
- Location of the horizontal stabilizer, if any.
- Dihedral angle – positive, zero, or negative (anhedral).

A variable geometry aircraft can change its wing configuration during flight.

A *flying wing* has no fuselage, though it may have small blisters or pods. The opposite of this is a *lifting body*, which has no wings, though it may have small stabilizing and control surfaces.

Wing-in-ground-effect vehicles are not considered aircraft. They "fly" efficiently close to the surface of the ground or water, like conventional aircraft during takeoff. An example is the Russian ekranoplan (nicknamed the "Caspian Sea Monster"). Man-powered aircraft also rely on ground effect to remain airborne with a minimal pilot power, but this is only because they are so under-powered—in fact, the airframe is capable of flying higher.

## Rotorcraft

An autogyro

Rotorcraft, or rotary-wing aircraft, use a spinning rotor with aerofoil section blades (a *rotary wing*) to provide lift. Types include helicopters, autogyros, and various hybrids such as gyrodynes and compound rotorcraft.

*Helicopters* have a rotor turned by an engine-driven shaft. The rotor pushes air downward to create lift. By tilting the rotor forward, the downward flow is tilted backward, producing thrust for forward flight. Some helicopters have more than one rotor and a few have rotors turned by gas jets at the tips.

*Autogyros* have unpowered rotors, with a separate power plant to provide thrust. The rotor is tilted backward. As the autogyro moves forward, air blows upward across the rotor, making it spin. This spinning increases the speed of airflow over the rotor, to provide lift. Rotor kites are unpowered autogyros, which are towed to give them forward speed or tethered to a static anchor in high-wind for kited flight.

*Cyclogyros* rotate their wings about a horizontal axis.

*Compound rotorcraft* have wings that provide some or all of the lift in forward flight. They are nowadays classified as *powered lift* types and not as rotorcraft. *Tiltrotor* aircraft (such as the V-22 Osprey), tiltwing, tailsitter, and coleopter aircraft have their rotors/propellers horizontal for vertical flight and vertical for forward flight.

## Other Methods of Lift

- A *lifting body* is an aircraft body shaped to produce lift. If there are any wings, they are too small to provide significant lift and are used only for stability and control. Lifting bodies are not efficient: they suffer from high drag, and must also travel at high speed to generate enough lift to fly. Many of the research prototypes, such as the Martin-Marietta X-24, which led up to the Space Shuttle, were lifting bodies (though the shuttle itself is not), and some supersonic missiles obtain lift from the airflow over a tubular body.

- *Powered lift* types rely on engine-derived lift for vertical takeoff and landing (VTOL). Most types transition to fixed-wing lift for horizontal flight. Classes of powered lift types include VTOL jet aircraft (such as the Harrier jump-jet) and tiltrotors (such as the V-22 Osprey), among others. A few experimental designs rely entirely on engine thrust to provide lift throughout the whole flight, including personal fan-lift hover platforms and jetpacks. VTOL research designs include the flying Bedstead.
- The *Flettner airplane* uses a rotating cylinder in place of a fixed wing, obtaining lift from the magnus effect.
- The *ornithopter* obtains thrust by flapping its wings.

X-24B lifting body, specialized glider

## Scale, Sizes and Speeds

### Sizes

The smallest aircraft are toys, and—even smaller -- nano-aircraft.

The largest aircraft by dimensions and volume (as of 2016) is the 302-foot-long (about 95 meters) British Airlander 10, a hybrid blimp, with helicopter and fixed-wing features, and reportedly capable of speeds up to 90 mph (about 150 km/h), and an airborne endurance of two weeks with a payload of up to 22,050 pounds (11 tons).

The largest aircraft by weight and largest regular fixed-wing aircraft ever built (as of 2016), is the Antonov An-225. That Ukrainian-built 6-engine Russian transport of the 1980s is 84 meters (276 feet) long, with an 88-meter (289 foot) wingspan. It holds the world payload record, after transporting 428,834 pounds (200 tons) of goods, and has recently flown 100-ton loads commercially. Weighing in at somewhere between 1.1 and 1.4 million pounds (550-700 tons) maximum loaded weight, it is also the heaviest aircraft to be built, to date. It can cruise at 500 mph.

The largest military airplanes are the Ukrainian/Russian Antonov An-124 (world's second-largest airplane, also used as a civilian transport), and American Lockheed C-5 Galaxy transport, weighing, loaded, over 765,000 pounds (over 380 tons). The 8-engine, piston/propeller Hughes HK-1 "Spruce Goose," an American World War II wooden flying boat transport—with a greater wingspan (94 meters / 260 feet) than any current aircraft, and a tail-height equal to the tallest (Airbus A380-800 at 24.1 meters / 78 feet) -- flew only one short hop in the late 1940s, and never flew out of ground effect.

The largest civilian airplanes, apart from the above-noted An-225 and An-124, are the French Airbus Beluga cargo transport derivative of the Airbus A300 jet airliner, the American Boeing Dreamlifter cargo transport derivative of the Boeing 747 jet airliner/transport (the 747-200B was, at its creation in the 1960s, the heaviest aircraft ever built, with a maximum weight of 836,000 pounds (over 400 tons)), and the double-decker French Airbus A380 "super-jumbo" jet airliner (the world's largest passenger airliner).

## Speeds

The fastest recorded powered aircraft flight and fastest recorded aircraft flight of an air-breathing powered aircraft was of the NASA X-43A Pegasus, a scramjet-powered, hypersonic, lifting body experimental research aircraft, at Mach 9.6 (nearly 7,000 mph). The X-43A set that new mark, and broke its own world record (of Mach 6.3, nearly 5,000 mph, set in March, 2004) on its third and final flight on Nov. 16, 2004.

Prior to the X-43A, the fastest recorded powered airplane flight (and still the record for the fastest manned, powered airplane / fastest manned, non-spacecraft aircraft) was of the North American X-15A-2, rocket-powered airplane at 4,520 mph (7,274 km/h), Mach 6.72, on October 3, 1967. On one flight it reached an altitude of 354,300 feet.

The fastest known, production aircraft (other than rockets and missiles) currently or formerly operational (as of 2016) are:

- The fastest fixed-wing aircraft, and fastest glider, is the Space Shuttle, a rocket-glider hybrid, which has re-entered the atmosphere as a fixed-wing glider at over Mach 25 (over 25 times the speed of sound—about 17,000 mph at re-entry to Earth's atmosphere).
- The fastest military airplane ever built: Lockheed SR-71 Blackbird, a U.S. reconnaissance jet fixed-wing aircraft, known to fly beyond Mach 3.3 (about 2,200 mph at cruising altitude). On July 28, 1976, an SR-71 set the record for the fastest and highest-flying operational aircraft with an absolute speed record of 2,193 mph and an absolute altitude record of 85,068 feet. At its retirement in the January 1990, it was the fastest air-breathing aircraft / fastest jet aircraft in the world—a record still standing as of August, 2016.

  Note: Some sources refer to the above-mentioned X-15 as the "fastest military airplane" because it was partly a project of the U.S. Navy and Air Force; however, the X-15 was not used in non-experimental actual military operations.
- The fastest current military aircraft are the Soviet/Russian MiG-25—capable of Mach 3.2 (2,170 mph), at the expense of engine damage, or Mach 2.83 (1,920 mph) normally—and the Russian MiG-31E (also capable of Mach 2.83 normally). Both are fighter-interceptor jet airplanes, in active operations as of 2016.

- The fastest civilian airplane ever built, and fastest passenger airliner ever built: the briefly operated Tupolev Tu-144 supersonic jet airliner (Mach 2.35, 1,600 mph, 2,587 km/h), which was believed to cruise at about Mach 2.2. The Tu-144 (officially operated from 1968 to 1978, ending after two crashes of the small fleet) was outlived by its rival, the Concorde SST (Mach 2.23), a French/British supersonic airliner, known to cruise at Mach 2.02 (1.450 mph, 2,333 kmh at cruising altitude), operating from 1976 until the small Concorde fleet was grounded permanently in 2003, following the crash of one in the early 2000s.
- The fastest civilian airplane currently flying: the Cessna Citation Ten, an American business jet, capable of Mach 0.935 (over 600 mph at cruising altitude). Its rival, the American Gulfstream 650 business jet, can reach Mach 0.925
- The fastest airliner currently flying is the Boeing 747, quoted as being capable of cruising over Mach 0.885 (over 550 mph). Previously, the fastest were the troubled, short-lived Russian (Soviet Union) Tupolev Tu-144 SST (Mach 2.35) and the French/British Concorde SST (Mach 2.23, normally cruising at Mach 2) . Before them, the Convair 990 Coronado jet airliner of the 1960s flew at over 600 mph.

## Propulsion

### Unpowered Aircraft

Gliders are heavier-than-air aircraft that do not employ propulsion once airborne. Take-off may be by launching forward and downward from a high location, or by pulling into the air on a tow-line, either by a ground-based winch or vehicle, or by a powered "tug" aircraft. For a glider to maintain its forward air speed and lift, it must descend in relation to the air (but not necessarily in relation to the ground). Many gliders can 'soar' – gain height from updrafts such as thermal currents. The first practical, controllable example was designed and built by the British scientist and pioneer George Cayley, whom many recognise as the first aeronautical engineer. Common examples of gliders are sailplanes, hang gliders and paragliders.

Balloons drift with the wind, though normally the pilot can control the altitude, either by heating the air or by releasing ballast, giving some directional control (since the wind direction changes with altitude). A wing-shaped hybrid balloon can glide directionally when rising or falling; but a spherically shaped balloon does not have such directional control.

Kites are aircraft that are tethered to the ground or other object (fixed or mobile) that maintains tension in the tether or kite line; they rely on virtual or real wind blowing over and under them to generate lift and drag. Kytoons are balloon-kite hybrids that are shaped and tethered to obtain kiting deflections, and can be lighter-than-air, neutrally buoyant, or heavier-than-air.

### Powered Aircraft

Powered aircraft have one or more onboard sources of mechanical power, typically aircraft engines although rubber and manpower have also been used. Most aircraft engines are either lightweight piston engines or gas turbines. Engine fuel is stored in tanks, usually in the wings but larger aircraft also have additional fuel tanks in the fuselage.

## Propeller Aircraft

A turboprop-engined DeHavilland Twin Otter adapted as a floatplane

Propeller aircraft use one or more propellers (airscrews) to create thrust in a forward direction. The propeller is usually mounted in front of the power source in *tractor configuration* but can be mounted behind in *pusher configuration*. Variations of propeller layout include *contra-rotating propellers* and *ducted fans*.

Many kinds of power plant have been used to drive propellers. Early airships used man power or steam engines. The more practical internal combustion piston engine was used for virtually all fixed-wing aircraft until World War II and is still used in many smaller aircraft. Some types use turbine engines to drive a propeller in the form of a turboprop or propfan. Human-powered flight has been achieved, but has not become a practical means of transport. Unmanned aircraft and models have also used power sources such as electric motors and rubber bands.

## Jet Aircraft

Lockheed Martin F-22A Raptor

Jet aircraft use airbreathing jet engines, which take in air, burn fuel with it in a combustion chamber, and accelerate the exhaust rearwards to provide thrust.

Turbojet and turbofan engines use a spinning turbine to drive one or more fans, which provide additional thrust. An afterburner may be used to inject extra fuel into the hot exhaust, especially on military "fast jets". Use of a turbine is not absolutely necessary: other designs include the pulse jet and ramjet. These mechanically simple designs cannot work when stationary, so the aircraft must be launched to flying speed by some other method. Other variants have also been used, including the motorjet and hybrids such as the Pratt & Whitney J58, which can convert between turbojet and ramjet operation.

Compared to propellers, jet engines can provide much higher thrust, higher speeds and, above about 40,000 ft (12,000 m), greater efficiency. They are also much more fuel-efficient than rockets. As a consequence nearly all large, high-speed or high-altitude aircraft use jet engines.

### Rotorcraft

Some rotorcraft, such as helicopters, have a powered rotary wing or *rotor*, where the rotor disc can be angled slightly forward so that a proportion of its lift is directed forwards. The rotor may, like a propeller, be powered by a variety of methods such as a piston engine or turbine. Experiments have also used jet nozzles at the rotor blade tips.

### Other Types of Powered Aircraft

- *Rocket-powered aircraft* have occasionally been experimented with, and the Messerschmitt *Komet* fighter even saw action in the Second World War. Since then, they have been restricted to research aircraft, such as the North American X-15, which traveled up into space where air-breathing engines cannot work (rockets carry their own oxidant). Rockets have more often been used as a supplement to the main power plant, typically for the rocket-assisted take off of heavily loaded aircraft, but also to provide high-speed dash capability in some hybrid designs such as the Saunders-Roe SR.53.
- The *ornithopter* obtains thrust by flapping its wings. It has found practical use in a model hawk used to freeze prey animals into stillness so that they can be captured, and in toy birds.

### Design and Construction

Aircraft are designed according to many factors such as customer and manufacturer demand, safety protocols and physical and economic constraints. For many types of aircraft the design process is regulated by national airworthiness authorities.

The key parts of an aircraft are generally divided into three categories:

- The *structure* comprises the main load-bearing elements and associated equipment.
- The *propulsion system* (if it is powered) comprises the power source and associated equipment, as described above.
- The *avionics* comprise the control, navigation and communication systems, usually electrical in nature.

### Structure

The approach to structural design varies widely between different types of aircraft. Some, such as paragliders, comprise only flexible materials that act in tension and rely on aerodynamic pressure to hold their shape. A balloon similarly relies on internal gas pressure but may have a rigid basket or gondola slung below it to carry its payload. Early aircraft, including airships, often employed flexible doped aircraft fabric covering to give a reasonably smooth aeroshell stretched over a rigid frame. Later aircraft employed semi-monocoque techniques, where the skin of the aircraft is stiff enough to share much of the flight loads. In a true monocoque design there is no internal structure left.

The key structural parts of an aircraft depend on what type it is.

## Aerostats

Lighter-than-air types are characterised by one or more gasbags, typically with a supporting structure of flexible cables or a rigid framework called its hull. Other elements such as engines or a gondola may also be attached to the supporting structure.

## Aerodynes

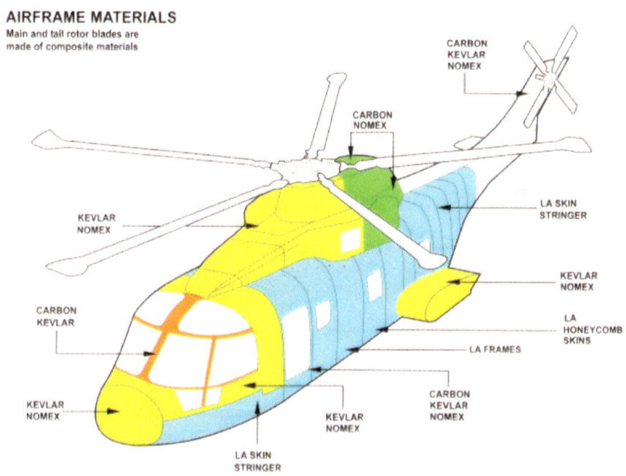


Airframe diagram for an AgustaWestland AW101 helicopter

Heavier-than-air types are characterised by one or more wings and a central fuselage. The fuselage typically also carries a tail or empennage for stability and control, and an undercarriage for takeoff and landing. Engines may be located on the fuselage or wings. On a fixed-wing aircraft the wings are rigidly attached to the fuselage, while on a rotorcraft the wings are attached to a rotating vertical shaft. Smaller designs sometimes use flexible materials for part or all of the structure, held in place either by a rigid frame or by air pressure. The fixed parts of the structure comprise the airframe.

## Avionics

The avionics comprise the flight control systems and related equipment, including the cockpit instrumentation, navigation, radar, monitoring, and communication systems.

## Flight Characteristics

## Flight Envelope

The flight envelope of an aircraft refers to its capabilities in terms of airspeed and load factor or altitude. The term can also refer to other measurements such as maneuverability. When a craft is pushed, for instance by diving it at high speeds, it is said to be flown *outside the envelope*, something considered unsafe.

## Range

The Boeing 777-200LR is the longest-range airliner, capable of flights of more than halfway around the world.

The range is the distance an aircraft can fly between takeoff and landing, as limited by the time it can remain airborne.

For a powered aircraft the time limit is determined by the fuel load and rate of consumption.

For an unpowered aircraft, the maximum flight time is limited by factors such as weather conditions and pilot endurance. Many aircraft types are restricted to daylight hours, while balloons are limited by their supply of lifting gas. The range can be seen as the average ground speed multiplied by the maximum time in the air.

## Flight Dynamics

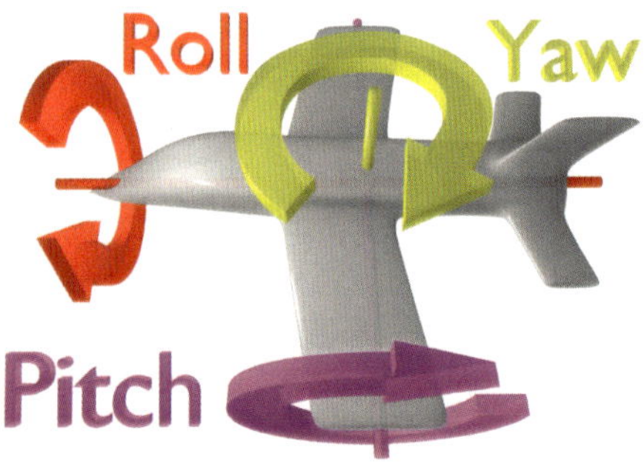


Flight dynamics is the science of air vehicle orientation and control in three dimensions. The three critical flight dynamics parameters are the angles of rotation around three axes about the vehicle's center of mass, known as *pitch*, *roll*, and *yaw* (quite different from their use as Tait-Bryan angles).

- Roll is a rotation about the longitudinal axis (equivalent to the rolling or heeling of a ship) giving an up-down movement of the wing tips measured by the roll or bank angle.
- Pitch is a rotation about the sideways horizontal axis giving an up-down movement of the aircraft nose measured by the angle of attack.
- Yaw is a rotation about the vertical axis giving a side-to-side movement of the nose known as sideslip.

Flight dynamics is concerned with the stability and control of an aircraft's rotation about each of these axes.

## Stability

The tail assembly of a Boeing 747–200

An aircraft that is unstable tends to diverge from its current flight path and so is difficult to fly. A very stable aircraft tends to stay on its current flight path and is difficult to manoeuvre—so it is important for any design to achieve the desired degree of stability. Since the widespread use of digital computers, it is increasingly common for designs to be inherently unstable and rely on computerised control systems to provide artificial stability.

A fixed wing is typically unstable in pitch, roll, and yaw. Pitch and yaw stabilities of conventional fixed wing designs require horizontal and vertical stabilisers, which act similarly to the feathers on an arrow. These stabilizing surfaces allow equilibrium of aerodynamic forces and to stabilise the flight dynamics of pitch and yaw. They are usually mounted on the tail section (empennage), although in the canard layout, the main aft wing replaces the canard foreplane as pitch stabilizer. Tandem wing and Tailless aircraft rely on the same general rule to achieve stability, the aft surface being the stabilising one.

A rotary wing is typically unstable in yaw, requiring a vertical stabiliser.

A balloon is typically very stable in pitch and roll due to the way the payload is hung underneath.

## Control

Flight control surfaces enable the pilot to control an aircraft's flight attitude and are usually part of the wing or mounted on, or integral with, the associated stabilizing surface. Their development was a critical advance in the history of aircraft, which had until that point been uncontrollable in flight.

Aerospace engineers develop control systems for a vehicle's orientation (attitude) about its center of mass. The control systems include actuators, which exert forces in various directions, and generate rotational forces or moments about the aerodynamic center of the aircraft, and thus rotate the aircraft in pitch, roll, or yaw. For example, a pitching moment is a vertical force applied at a distance forward or aft from the aerodynamic center of the aircraft, causing the aircraft to pitch up or down. Control systems are also sometimes used to increase or decrease drag, for example to slow the aircraft to a safe speed for landing.

The two main aerodynamic forces acting on any aircraft are lift supporting it in the air and drag opposing its motion. Control surfaces or other techniques may also be used to affect these forces directly, without inducing any rotation.

## Uses for Aircraft

Aircraft are produced in several different types optimized for various uses; military aircraft, which includes not just combat types but many types of supporting aircraft, and civil aircraft, which include all non-military types, experimental and model.

### Military

Boeing B-17E in flight

A military aircraft is any aircraft that is operated by a legal or insurrectionary armed service of any type. Military aircraft can be either combat or non-combat:

- Combat aircraft are aircraft designed to destroy enemy equipment using its own armament. Combat aircraft divide broadly into fighters and bombers, with several in-between types such as fighter-bombers and ground-attack aircraft (including attack helicopters).
- Non-combat aircraft are not designed for combat as their primary function, but may carry weapons for self-defense. Non-combat roles include search and rescue, reconnaissance, observation, transport, training, and aerial refueling. These aircraft are often variants of civil aircraft.

Most military aircraft are powered heavier-than-air types. Other types such as gliders and balloons have also been used as military aircraft; for example, balloons were used for observation during the American Civil War and World War I, and military gliders were used during World War II to land troops.

### Civil

Civil aircraft divide into *commercial* and *general* types, however there are some overlaps.

Commercial aircraft include types designed for scheduled and charter airline flights, carrying passengers, mail and other cargo. The larger passenger-carrying types are the airliners, the largest of which are wide-body aircraft. Some of the smaller types are also used in general aviation, and some of the larger types are used as VIP aircraft.

Agusta A109 helicopter of the Swiss air rescue service

General aviation is a catch-all covering other kinds of private (where the pilot is not paid for time or expenses) and commercial use, and involving a wide range of aircraft types such as business jets (bizjets), trainers, homebuilt, gliders, warbirds and hot air balloons to name a few. The vast majority of aircraft today are general aviation types.

### Experimental

An experimental aircraft is one that has not been fully proven in flight, or that carries an FAA airworthiness certificate in the "Experimental" category. Often, this implies that the aircraft is testing new aerospace technologies, though the term also refers to amateur and kit-built aircraft—many based on proven designs.

### Model

A model aircraft, weighing six grams

A model aircraft is a small unmanned type made to fly for fun, for static display, for aerodynamic research or for other purposes. A scale model is a replica of some larger design.

## Airliner

Airliners are planes that carry paying passengers along regular routes. Airline services began after World War I (1914–1918). Some of the earliest airliners were bomber planes converted to carry people, not bombs. Most were made largely of wood and fabric. By the mid-1920s, there were many specially made airliners operating all over the world. In the 1930s, America's Boeing 247, the Douglas DC-2 and the Douglas DC-3 revolutionized air travel. These sleek, all-metal airliners could cover large distances at impressive speeds. They helped show that air travel could be safe, comfortable, and reliable, even though it remained expensive.

## Some Examples

### Boeing Jet Airliners

The success of the 707 made Boeing the leader in commercial airliners, and led to a popular family of jetliners introduced over the years: the 717, 727, 737, 747, 757, 767, 777, and the 787.

One can spot a Boeing jet airliner in nearly any airport in the world. They share similar external characteristics, but identifying one from another can be difficult.

### Airbus Airliners

The A300 was the world's first twin-engined widebody airliner, and the first produced by Airbus. It was stretched into the A330 and A340, and shortened to the A310.

The popular Airbus A320 is a short-to-medium range, twin-engine airliner. The A350 XWB is now in service, and the A380 is the world's largest commercial aircraft.

## Bombardier and Embraer

The aviation industry manufactures and operates a large number of regional and mid-range jetliners.

Popular models include the Bombardier CRJ and C Series, and the Embraer ERJ and E Series jets.

## Types

## Wide-body Airliners

Boeing 747-8I wide-body airliner, first flight

The largest airliners are *wide-body* jets. These aircraft are frequently called *twin-aisle aircraft* because they generally have two separate aisles running from the front to the back of the passenger cabin. Aircraft in this category are the Boeing 747, Boeing 767, Boeing 777, Boeing 787, Airbus A300/A310, Airbus A330, Airbus A340, Airbus A350, Airbus A380, Lockheed L-1011 TriStar, McDonnell Douglas DC-10, McDonnell Douglas MD-11, Ilyushin Il-86, and Ilyushin Il-96. These aircraft are usually used for long-haul flights between airline hubs and major cities with many passengers.

## Narrow-body Airliners

A smaller, more common class of airliners is the *narrow-body* or *single aisle* aircraft. These smaller airliners are generally used for medium haul flights with fewer passengers than their wide-body counterparts. Currently produced narrow-body airliners are the Boeing 737, Airbus A320 family, Bombardier CSeries or Tupolev Tu-204/214.

Airbus A321, a narrow-body airliner

Out-of production aircraft include the McDonnell Douglas DC-9 and its derivatives MD-80/MD-90 series and Boeing 717, using the same cabin cross section as the Boeing 737 the 757, 727 and the 707, its competitor the Douglas DC-8 or the Tupolev Tu-154, Ilyushin IL-18, and the Ilyushin IL-62. Future narrow-body airliners include the Comac C919 and Irkut MC-21.

## Regional, Short Haul and Feederliner Aircraft

JetBlue Airways Embraer 190, short haul airliner

PLUNA Bombardier CRJ900, short haul (regional) airliner taxiing

Direktflyg Jetstream 32, Kristiansund Airport, Kvernberget

*Regional airliners* typically seat fewer than 100 passengers and may be powered by turbofans or turboprops. These airliners are the non-mainline counterparts to the larger aircraft operated by the major carriers, legacy carriers, and flag carriers and are used to feed traffic into the large airline hubs or focus cities. These particular routes may need the size of a smaller aircraft to meet the frequency needs and service levels customers expect in the marketed product that is offered by larger airlines and their modern narrow and widebody aircraft. Therefore, these short-haul airliners are usually equipped with lavatories, stand up cabins, pressurization, overhead storage bins, reclining seats, and have a flight attendant to look after the in-flight needs of the passengers during Point-to-point transit or routes.

Because these aircraft are frequently operated by smaller airlines that are contracted to provide ("feed") passengers from smaller cities to hub airports (and reverse) for a "major" or "flag" carrier, regional airliners may be painted in the liveries of the major airline for whom they provide this "feeder" service so the regional airlines may offer and market a seamless transition between the larger airline to smaller airline.

Bombardier CRJ200

Typical aircraft in this category include the Bombardier CRJ and Embraer ERJ regional jets along with the Bombardier "Q" (DASH-8) series, ATR 42/72 and Saab 340/2000 turboprop airliners.

## Commuterliners used by Regional Airlines and Air Taxi Operators

Beechcraft 1900, short range commuter aircraft

Compass Airlines (North America) Embraer ERJ-170-200LR in the feederliner colors of Northwest Airlink

The lightest (light aircraft, list of light transport aircraft) of short haul regional feeder airliner type aircraft that carry 19 or fewer passenger seats are called *commuter aircraft, commuterliners, feederliners,* and *air taxis*, depending on their size, engines, how they are marketed, region of the world, and seating configurations. The Beechcraft 1900, for example, has only 19 seats. Depending on local and national regulations, a commuter aircraft may not qualify as an airliner and may not be subject to the regulations applied to larger aircraft. Members of this class of aircraft normally lack such amenities as lavatories and galleys and typically do not carry a flight attendant as an aircrew member.

Other aircraft that may fall into this category are the Fairchild Metro, Jetstream 31, and Embraer EMB 110 Bandeirante. The Cessna Caravan and Pilatus PC-12 are single-engine turboprops, sometimes used as small airliners, although many countries stipulate a minimum requirement of two engines for aircraft to be used as airliners.

Twin piston-engined aircraft made by Cessna, Piper, Britten-Norman, and Beechcraft are also in use as short haul, short range commuter type aircraft.

## Engines

Until the beginning of the Jet Age, piston engines were common on propliners like the Douglas DC-3. Nearly all modern airliners are now powered by turbine engines, either turbofans or turboprops. Gas turbine engines operate efficiently at much higher altitudes, are more reliable than piston engines, and produce less vibration and noise. The use of a common fuel type – kerosene-based jet fuel – is another advantage. Prior to the Jet Age, it was common for the same or very similar engines to be used in civilian airliners as in military aircraft. In recent years, divergence has occurred so that it is now unusual for the same engine to be used on a military type and a civilian type. Those military aircraft which do share engine technology with airliners are typically transports or tanker types

## Configuration

Modern jetliners are usually low-wing designs with two engines mounted underneath the swept wings (turboprop aircraft are slow enough to use straight wings). The Boeing 747 and Airbus A380 are the only airliners in production which are too heavy (more than 400 tons maximum takeoff weight) for just two engines. Smaller airliners sometimes have their engines mounted on either side of the rear fuselage. There are numerous advantages and disadvantages to this arrangement. Perhaps the most important advantage to mounting the engines under the wings is that the total aircraft weight is more evenly distributed across the wingspan, which imposes less bending moment on the wings and allows for a lighter wing structure. This factor becomes more important as aircraft weight increases, and there are no in-production airliners with both a maximum takeoff weight of more than 50 tons and engines mounted on the fuselage. The Antonov An-148 is the only in-production jetliner with high-mounted wings (usually seen in military transport aircraft), which reduces the risk of damage from unpaved runways.

Except for a few experimental or military designs, all aircraft built to date have had all of their weight lifted off the ground by airflow across the wings. In terms of aerodynamics, the fuselage has been a mere burden. NASA and Boeing are currently developing a blended wing body design in which the entire airframe, from wingtip to wingtip, contributes lift. This promises a significant gain in fuel efficiency.

## Current Manufacturers

Assembly of a Boeing 767 nose section

The major manufacturers with airliners currently in production include:

- Airbus (France/Germany/Spain/United Kingdom)
- Antonov (Ukraine)
- ATR Aircraft (France/Italy)
- Boeing (United States)
- Bombardier Aerospace (Canada)
- Comac (China)
- Embraer (Brazil)
- Let Kunovice (Czech Republic)
- Mitsubishi Aircraft Corporation (Japan)
- United Aircraft Corporation (Russia, includes Tupolev, Yakovlev, Sukhoi, Ilyushin, and others)
- Xi'an Aircraft Industrial Corporation (China)

The narrowbody and widebody airliner market is dominated by Airbus and Boeing, and the regional airliner market is mostly split between ATR Aircraft, Bombardier Aerospace and Embraer.

## Notable Airliners

- Boeing 247 – the first design to incorporate modern features such as all-metal construction and retractable landing gear
- Douglas DC-3 – still in service more than 80 years after its debut, it is generally regarded as one of the most significant transport aircraft ever made
- Boeing 307 – one of the first airliners to be pressurized and have the ability to fly into the stratosphere

Airbus A380 "superjumbo" of Singapore Airlines takes off

Antonov An-225 *Mriya* "Mriya"

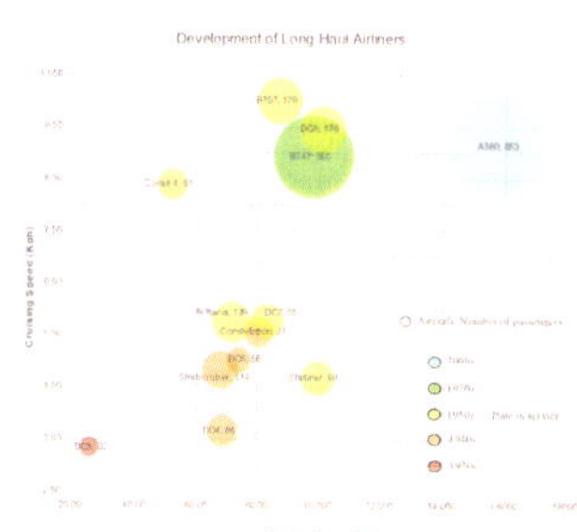

Development of the capabilities of long haul airliners shown by some notable ones

- Douglas DC-6 – originally developed as a military transport, it was reworked for passenger service after World War II, a role it continues to perform today.

- Boeing 377- developed soon after World War 2, from the C-97 Stratofreighter, this was a luxurious double-decker airliner with a pressurized cabin.
- Vickers Viscount – the first turboprop airliner to enter service.
- Lockheed Constellation – a distinctive triple-tailed piston-engined airliner of the 1950s, it was one of the last large propeller-driven airliners.
- De Havilland Comet – the world's first jetliner to reach mass production, early crashes informed vital design lessons.
- Antonov An-2- best selling transport airliner up to the point it was built.
- Tupolev Tu-104 – the first turbojet airliner to provide sustained service, and the sole jetliner operating in the world between 1956 and 1958.
- Boeing 707 – the first United States-built jetliner to enter production.
- Tupolev Tu-124 – the first turbo-fan powered airliner.
- Douglas DC-8 – launched after the Boeing 707, it nevertheless established Douglas in the airliner market, and continues to serve as a cargo aircraft to this day.
- Sud Aviation Caravelle – the first jetliner with rear mounted engines. It is also the first airliner to perform a CAT III landing (low visibility).
- Tupolev Tu-154 – standard medium-range airliner for Russia (and others), carried half of all Soviet air traffic since 1972 with 1015 built; the fastest airliner in service.
- Ilyushin Il-62 – standard long-range airliner for Russia (and others) for three decades, first flight 1963; still in service.
- Douglas DC-9 – production of it and successive variants nearly reached 2,500.
- Boeing 737 – the best-selling jet airliner in the history of aviation.
- Hawker Siddeley Trident The first airliner to make an automated blind landing whilst in service.
- Tupolev Tu-144 – the first supersonic transport aircraft constructed in the Soviet Union.
- Concorde – an Anglo-French supersonic transport, it remains the only supersonic aircraft to sustain a regular passenger service.
- Boeing 747 "Jumbo Jet" – an iconic aircraft, it was the world's largest passenger airliner between 1969 and 2005. It was also the first widebody jet airliner to enter service, introducing many features that became standard in current jetliners like high-bypass turbofans.
- McDonnell Douglas DC-10 – a trijet competitor to the widebody 747.
- Lockheed L-1011 TriStar – a trijet competitor to the Boeing 747 and the similarly sized DC-10. Lockheed's last entry into commercial aircraft manufacturing.
- Airbus A300 – the world's first twinjet widebody.
- Airbus A320 – pioneered the use of fly-by-wire technology.

- Antonov An-225 *Mriya* "Mriya" – the world's largest and heaviest aircraft. Only one example built, it is used for carrying oversized payloads. It is also the only airliner with six engines.
- Boeing 777 – the first airliner designed entirely by computer, without physical mockups.
- Airbus A380 "Superjumbo" – the world's largest passenger airliner from 2005 onwards.
- Boeing 787 Dreamliner – the world's first jet airliner to make use of composite materials for most of its construction.

## Cabin Configurations and Features

Interior of a Qatar Airways Airbus. Video systems (the vertical white panels) are visible above the very centre seats of the aircraft

An airliner will usually have several classes of seating: first class, business class, and/or economy class (which may be referred to as coach class or tourist class, and sometimes has a separate "premium" economy section with more legroom and amenities). The seats in more expensive classes are wider, more comfortable, and have more amenities such as "lie flat" seats for more comfortable sleeping on long flights. Generally, the more expensive the class, the better the beverage and meal service.

Domestic flights generally have a two-class configuration, usually first or business class and coach class, although many airlines instead offer all-economy seating. International flights generally have either a two-class configuration or a three-class configuration, depending on the airline, route and aircraft type. Many airliners offer movies or audio/video on demand (this is standard in first and business class on many international flights and may be available on economy). Cabins of any class are provided with lavatory facilities, reading lights and gaspers. Larger airliners may have a segregated rest compartment reserved for crew use during breaks.

## Seats

The types of seats that are provided and how much legroom is given to each passenger are decisions made by the individual airlines, not the aircraft manufacturers. Seats are mounted in "tracks" on the floor of the cabin and can be moved back and forth by the maintenance staff or removed altogether. Naturally the airline tries to maximize the number of seats available in every aircraft to carry the largest possible (and therefore most profitable) number of passengers.

Passengers seated in an exit row (the row of seats adjacent to an emergency exit) usually have substantially more legroom than those seated in the remainder of the cabin, while the seats directly in front of the exit row may have less legroom and may not even recline (for evacuation safety reasons). However, passengers seated in an exit row may be required to assist cabin crew during an emergency evacuation of the aircraft opening the emergency exit and assisting fellow passengers to the exit. As a precaution, many airlines prohibit young people under the age of 15 from being seated in the exit row.

The seats are designed to withstand strong forces so as not to break or come loose from their floor tracks during turbulence or accidents. The backs of seats are often equipped with a fold-down tray for eating, writing, or as a place to set up a portable computer, or a music or video player. Seats without another row of seats in front of them have a tray that is either folded into the armrest or that clips into brackets on the underside of the armrests. However, seats in premium cabins generally have trays in the armrests or clip-on trays, regardless of whether there is another row of seats in front of them. Seatbacks now often feature small colour LCD screens for videos, television and video games. Controls for this display as well as an outlet to plug in audio headsets are normally found in the armrest of each seat.

## Overhead Bins

Overhead bins aboard a Sukhoi Superjet 100

The overhead bins are used for stowing carry-on baggage and other items. While the airliner manufacturer will normally specify a standard version of the product to supply, airlines can choose to have bins of differing size, shape, or color installed. Over time, overhead bins evolved out of what were originally overhead shelves that were used for little more than coat and briefcase storage. As concerns about falling debris during turbulence or in accidents increased, enclosed bins became the norm. Bins have increased in size to accommodate the larger carry-on baggage passengers can bring onto the aircraft. Newer bin designs have included a handrail, useful when moving through the cabin.

## Passenger Service Units

Above the passenger seats are Passenger Service Units (PSU). These typically contain reading lights, air vents, and a flight attendant call light. On most narrowbody aircraft (and some Airbus A300s and A310s), the flight attendant call button and the buttons to control the reading lights are located directly on the PSU, while on most widebody aircraft, the flight attendant call button and the reading light control buttons are usually part of the in-flight entertainment system. The units frequently have small "Fasten Seat Belt" and "No Smoking" illuminated signage and may also contain a speaker for the cabin public address system.

The PSU will also normally contain the drop-down oxygen masks which are activated if there is a sudden drop in cabin pressure. These are supplied with oxygen by means of a chemical oxygen generator. By using a chemical reaction rather than a connection to an oxygen tank, these devices supply breathing oxygen for long enough for the airliner to descend to thicker, more breathable air. Oxygen generators do generate considerable heat in the process. Because of this, the oxygen generators are thermally shielded and are only allowed in commercial airliners when properly installed – they are not permitted to be loaded as freight on passenger-carrying flights. ValuJet Flight 592 crashed on May 11, 1996, as a result of improperly loaded chemical oxygen generators.

## Cabin Pressurization

Airliners developed since the 1940s have had pressurized cabins (or, more accurately, pressurized hulls including baggage holds) to enable them to carry passengers safely at high altitudes where low oxygen levels and air pressure would otherwise cause sickness or death. High altitude flight enabled airliners to fly above most weather systems that cause turbulent or dangerous flying conditions, and also to fly faster and further as there is less drag due to the lower air density. Pressurization is applied using compressed air, in most cases bled from the engines, and is managed by an environmental control system which draws in clean air, and vents stale air out through a valve.

Pressurization presents design and construction challenges to maintain the structural integrity and sealing of the cabin and hull and to prevent rapid decompression. Some of the consequences include small round windows, doors that open inwards and are larger than the door hole, and an emergency oxygen system.

To maintain a pressure in the cabin equivalent to an altitude close to sea level would, at a cruising altitude around 10,000 m (33,000 ft), create a pressure difference between inside the aircraft and outside the aircraft that would require greater hull strength and weight. Most people do not suffer ill effects up to an altitude of 1,800–2,500 m (5,900–8,200 ft), and maintaining cabin pressure at this equivalent altitude significantly reduces the pressure difference and therefore the required hull strength and weight. A side effect is that passengers experience some discomfort as the cabin pressure changes during ascent and descent to the majority of airports, which are at low altitudes.

## Cabin Climate Control

The air bled from the engines is hot and requires cooling by air conditioning units. It is also extremely dry at cruising altitude, and this causes sore eyes, dry skin and mucosa on long flights. Although humidification technology could raise its relative humidity to comfortable middle levels, this is not done since humidity promotes corrosion to the inside of the hull and risks condensation which could short electrical systems, so for safety reasons it is deliberately kept to a low value, around 10%. Another problem of the air coming from the ventilation (unto which the oil lubrication system of the engines is hooked up) is that fumes from components in the synthetic oils can sometimes travel along, causing passengers, pilots and crew to be intoxicated. The illness it causes is called aerotoxic syndrome.

## Baggage Holds

Airbus A320 baggage hold

Loading luggage onto a Boeing 747 at Boston Logan Airport, during snow

Airliners must have space on board to store "checked" baggage – that which will not safely fit in the passenger cabin.

Designed to hold baggage as well as freight, these compartments are called "cargo bins", "holds", or occasionally "pits". Occasionally baggage holds may be referred to as cargo decks on the largest of aircraft. These compartments can be accessed through doors on the outside of the aircraft.

Boeing 747 front lower compartment. Note the rollers for ULDs on the floor and the partition labeled "Caution: Do Not Hit – Potable Water Tank Inside".

Depending on the aircraft, baggage holds are normally inside the hull and are therefore pressurized just like the passenger cabin although they may not be heated. While lighting is normally installed for use by the loading crew, typically the compartment is unlit when the door is closed.

Baggage holds on modern airliners are equipped with fire detection equipment and larger aircraft have automated or remotely activated fire-fighting devices installed.

## Narrow-body Airliners

Most "narrow-body" airliners with more than 100 seats have space below the cabin floor, while smaller aircraft often have a special compartment separate from the passenger area but on the same level.

Baggage is normally stacked within the bin by hand, sorted by destination category. Netting that fits across the width of the bin is secured to limit movement of the bags. Airliners often carry items of freight and mail. These may be loaded separately from the baggage or mixed in if they are bound for the same destination. For securing bulky items "hold down" rings are provided to tie items into place.

### Wide-body Airliners

"Wide-body" airliners frequently have a compartment like the ones described above, typically called a "bulk bin". It is normally used for late arriving luggage or bags which may have been checked at the gate.

However, most baggage and loose freight items are loaded into containers called Unit Load Devices (ULDs), often referred to as "cans". ULDs come in a variety of sizes and shapes, but the most common model is the LD3. This particular container has approximately the same height as the cargo compartment and fits across half of its width.

ULDs are loaded with baggage and are transported to the aircraft on dolly carts and loaded into the baggage hold by a loader designed for the task. By means of belts and rollers an operator can maneuver the ULD from the dolly cart, up to the aircraft baggage hold door, and into the aircraft. Inside the hold, the floor is also equipped with drive wheels and rollers that an operator inside can use to move the ULD properly into place. Locks in the floor are used to hold the ULD in place during flight.

For consolidated freight loads, like a pallet of boxes or an item too oddly shaped to fit into a container, flat metal pallets that resemble large baking sheets that are compatible with the loading equipment are used.

## Powered Aircraft

| Powered aircraft use the energy provided by engines for propulsion. Many aircraft also achieve lift in the same manner. | | |
|---|---|---|
| Propeller aircraft | Propeller aircraft use propellers (i.e. engines with vertically-mounted blades) for propulsion. The blades in the propeller forms a helical surface and, when spinning, "cuts" the air in front of the blades in a screwing fashion. This, in turn, generates a forward motion on the aircraft. The power from the engines is also used for controlling the speed of the aircraft, which is achieved by adjusting the propeller's rotational speed and/or its angle of attack. Typical propeller aircraft are propeller airplanes, which also use the forward motion generated by the propeller to lift the aircraft and sustain flight. Propellers are also a common source of propulsion used by airships, autogyros and gyrodynes, which don't need them for lifting but only for navigating. |  |

| | | |
|---|---|---|
| Jet aircraft | Jet aircraft use jet engines for propulsion[2]. Jet engines take air at the front of the engines, compresses it at intense pressure and release it through a narrow end at the back of the engines, thus generating a forward motion. The power from the engines is also used for controlling the speed of the aircraft. Typical jet aircraft are jet airplanes, which also use the forward motion generated by the engines to lift the aircraft and sustain flight.. |  |
| Rocket aircraft | Rocket-powered aircraft use rockets for propulsion. Rockets work with stored fluids, which are ignited and released through a narrow end at the back of the engine, thus generating a forward motion. Rockets are typically used in spacecrafts. |  |
| Helicopter | Helicopters use rotors (i.e. engines with rotating top-mounted blades) in order to achieve both lift and onward motion (i.e., to move forward, backwards, upwards, etc). | 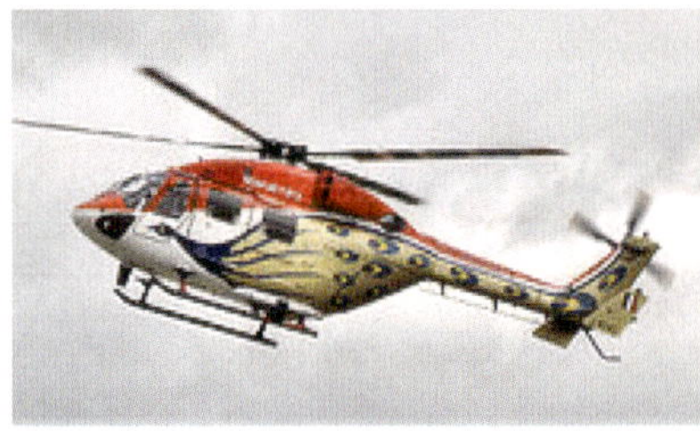 |

## Methods of Propulsion

## Rotating Aerofoils

Propeller Aircraft

A turboprop-engined Tupolev Tu-95

A propeller or airscrew comprises a set of small, wing-like aerofoil *blades* set around a central hub which spins on an axis aligned in the direction of travel. The blades are set at a *pitch* angle to the airflow, which may be fixed or variable, such that spinning the propeller creates aerodynamic lift, or *thrust*, in a forward direction.

A *tractor* design mounts the propeller in front of the power source, while a *pusher* design mounts it behind. Although the pusher design allows cleaner airflow over the wing, tractor configuration is more common because it allows cleaner airflow to the propeller and provides a better weight distribution.

*Contra-rotating propellers* have one propeller close behind another on the same axis, but rotating in the opposite direction.

A variation on the propeller is to use many broad blades to create a fan. Such fans are usually surrounded by a ring-shaped fairing or duct, as *ducted fans*.

Many kinds of power plant have been used to drive propellers.

Gossamer Albatross, a human-powered aircraft

The earliest designs used man power to give dirigible balloons some degree of control, and go back to Jean-Pierre Blanchard in 1784. Attempts to achieve heavier-than-air man-powered flight did not succeed fully until Paul MacCready's Gossamer Condor in 1977.

The first powered flight of an aircraft was made in a steam-powered dirigible by Henri Giffard in 1852. Attempts to marry a practical lightweight steam engine to a practical fixed-wing airframe did not succeed until much later, by which time the internal combustion engine was already dominant.

From the first controlled powered fixed-wing aircraft flight by the Wright brothers until World War II, propellers turned by the internal combustion piston engine were virtually the only type of propulsion system in use. The piston engine is still used in the majority of smaller aircraft produced, since it is efficient at the lower altitudes and slower speeds suited to propellers.

Turbine engines need not be used as jets, but may be geared to drive a propeller in the form of a turboprop. Modern helicopters also typically use turbine engines to power the rotor. Turbines provide more power for less weight than piston engines, and are better suited to small-to-medium size aircraft or larger, slow-flying types. Some turboprop designs mount the propeller directly on an engine turbine shaft, and are called propfans.

Other less common power sources include:

- Electric motors, often linked to solar panels to create a solar-powered aircraft.
- Rubber bands, wound many times to store energy, are mostly used for flying models.

## Rotorcraft

Rotorcraft have spinning blades called a *rotor* which spins in the horizontal plane to provide lift. Forward thrust is usually obtained by angling the rotor disc slightly forward so that a proportion of its lift is directed backwards; these are called helicopters. Other rotorcraft are compound helicopters and autogyros which sometimes use other means of propulsion, such as propellers and jets.

The rotor of a helicopter may, like a propeller, be powered by a variety of methods such as an internal-combustion engine or jet turbine. Tip jets, fed by gases passing along hollow rotor blades from a centrally mounted engine, have been experimented with. Attempts have even been made to mount engines directly on the rotor tips.

## Jet Propulsion

## Jet Aircraft

Airbreathing jet engines provide thrust by taking in air, compressing the air, injecting fuel into the hot compressed air mixture in a combustion chamber, the resulting accelerated exhaust ejects rearwards through a turbine which drives the compressor. The reaction against this acceleration provides the engine thrust.

Jet engines can provide much higher thrust than propellers, and are naturally efficient at higher altitudes, being able to operate above 40,000 ft (12,000 m). They are also much more fuel-efficient at normal flight speeds than rockets. Consequently, nearly all high-speed and high-altitude aircraft use jet engines.

A jet-engined Boeing 777 taking off

The early turbojet and modern turbofan use a spinning compressor and turbine to provide thrust. Many, mostly in military aviation, add an afterburner which injects extra fuel into the hot exhaust.

Use of a turbine is not absolutely necessary: other designs include the crude pulse jet, high-speed ramjet and the still-experimental supersonic-combustion ramjet or scramjet. These mechanically simple designs require an existing airflow to work and cannot work when stationary, so they must be launched by a catapult or rocket booster, or dropped from a mother ship.

The bypass turbofan engines of the Lockheed SR-71 were a hybrid design – the aircraft took off and landed in jet turbine configuration, and for high-speed flight the afterburner was lit and the turbine bypassed, to create a ramjet.

The motorjet was a very early design which used a piston engine in place of the combustion chamber, similar to a turbocharged piston engine except that the thrust is derived from the turbine instead of the crankshaft. It was soon superseded by the turbojet and remained a curiosity.

## Rocket-powered Aircraft

Rocket propulsion offers very high thrust for light weight and has no height limit, but suffers from high fuel consumption and the need to carry oxidant as well as propellant.

Rocket-powered aircraft have been experimented with, and during the Second World War the Messerschmitt *Komet* fighter was developed and used operationally. Since then they have been restricted to specialised niches, such as the Bell X-1 which broke the sound barrier or the North American X-15 which was capable of flying at extremely high altitudes at the border with space as it was not dependent on atmospheric oxygen.

Rockets have more often been used as a supplement to the main powerplant, typically in the case of rocket-assisted take off to give more power for a heavily loaded aircraft or reduce the takeoff run. In a number of designs such as the prototype "mixed-power" Saunders-Roe SR.53 interceptor a rocket was used to provide high-speed climb and speed to reach the target while a smaller turbojet provided a slower and more economical return to base.

## Ornithopter

The ornithopter obtains thrust by flapping its wings. When the wing flaps, as opposed to gliding, it continues to develop lift as before, but the lift is rotated forward to provide a thrust component.

Working devices have been created for flight research and as prototypes, but the vertical oscillation of the fuselage, which tends to accompany the wing flapping, limits their usefulness. The only practical application is a flying model hawk used to freeze prey animals into stillness so that they can be captured.

Toys in the form of a flying model bird are also popular.

## Methods of Powering Lift

A fixed-wing aircraft obtains lift from airflow over the wing resulting from motion due to forward thrust. A few other types, such as the rotary-winged autogyro, obtain lift through similar methods.

Some types use a separate power system to create lift. These include the rotary-winged helicopter and craft that use lift jets (e.g. the flying bedstead).

A hot air balloon requires a power source (normally a gas burner) for lift, but is not normally considered a "powered aircraft".

## Rotorcraft

A 'Rotorcraft' as an "aerodyne that derives the whole or a substantial part of its lift from a rotary wing system." An 'aerodyne' is "an aircraft, heavierthan-air, that can be dynamically sustained in the air by the reaction of the air upon surfaces in relative movement."

A 'helicopter' is defined by the FAI as "a rotorcraft which, in flight, derives substantially the whole of its lift from a power-driven rotor system whose axis (axes) is (are) fixed and substantially perpendicular to the longitudinal axis of the rotorcraft."

It is perhaps worthy of note that the tilt rotor configuration is specifically defined in its own right within the definition of Class E Rotorcraft. The convertible nature of this configuration clearly makes a strict comparison with rotorcraft configurations that cannot convert from one mode of flight to another virtually meaningless. This is a good example where records awarded to one configuration must be judged as recognition of endeavour that is different in a fundamental way from records awarded to another. It is for this reason that the FAI defines different types of aeroplanes and rotorcraft in the first place.

It is clearly important that the definitions used by the FAI properly and correctly differentiate between fundamentally different configurations of rotorcraft. The traditional helicopter derives both lift and longitudinal thrust from a power driven rotor that is substantially perpendicular to the axis of the rotorcraft. The FAI definition makes no mention of longitudinal thrust, only lift acting substantially perpendicular to the axis of the rotorcraft. It is a matter of great achievement that the edgewise rotor unsupported by fixed lifting surfaces and operating in the most complex aerodynamic conditions is able to achieve high speed. The speed record for a helicopter relying substantially on thrust generated by an edgewise rotor for lift and forward propulsion is an achievement to be uniquely celebrated and not confused with other rotorcraft configurations that may be equally worthy of reward but for different reasons.

## Classes of Rotorcraft

### Helicopter

A helicopter is a rotorcraft whose rotors are driven by the engine(s) throughout the flight to allow the helicopter to take off vertically, hover, fly forwards, backwards and laterally, as well as to land vertically. Helicopters have several different configurations of one or more main rotors.

Helicopters with a single shaft-driven main lift rotor require some sort of antitorque device such as a tail rotor, fantail, or NOTAR, except some rare examples of helicopters using tip jet propulsion, which generates almost no torque.

### Cyclogyro/Cyclocopter

A cyclocopter is a rotorcraft whose rotors are also driven by the engine throughout the flight, but the blades rotate about the horizontal axis while being parallel to it. They are being developed in a number of countries in order to replace the helicopters, which have a number of shortcomings such as low efficiency in forward flight, low speed, very high noise and vibration levels, limited flight range, and low altitude ceiling. At the present time flying model prototypes have been built in China, US, S. Korea and Austria.

### Autogyro

An autogyro (sometimes called gyrocopter, gyroplane, or rotaplane) utilizes an unpowered rotor, driven by aerodynamic forces in a state of autorotation to develop lift, and an engine-powered propeller, similar to that of a fixed-wing aircraft, to provide thrust. While similar to a helicopter rotor in appearance, the autogyro's rotor must have air flowing up and through the rotor disk in order to generate rotation. Early autogyros resembled the fixed-wing aircraft of the day, with wings and a front-mounted engine and propeller in a tractor configuration to pull the aircraft through the air. Late-model autogyros feature a rear-mounted engine and propeller in a pusher configuration.

A German-registered autogyro

The autogyro was invented in 1920 by Juan de la Cierva. The autogyro with pusher propeller was first tested by Etienne Dormoy with his Buhl A-1 Autogyro.

## Gyrodyne

The rotor of a gyrodyne is normally driven by its engine for takeoff and landing – hovering like a helicopter – with anti-torque and propulsion for forward flight provided by one or more propellers mounted on short or stub wings. As power is increased to the propeller, less power is required by the rotor to provide forward thrust resulting in reduced pitch angles and rotor blade flapping. At cruise speeds with most or all of the thrust being provided by the propellers, the rotor receives power only sufficient to overcome the profile drag and maintain lift. The effect is a rotorcraft operating in a more efficient manner than the freewheeling rotor of an autogyro in autorotation, minimizing the adverse effects of retreating blade stall of helicopters at higher airspeeds.

Fairey Rotodyne prototype

## Rotor Kite

A rotor kite or gyroglider is an unpowered rotary-wing aircraft. Like an autogyro or helicopter, it relies on lift created by one or more sets of rotors in order to fly. Unlike a helicopter, autogyros and rotor kites do not have an engine powering their rotors, but while an autogyro has an engine providing forward thrust that keeps the rotor turning, a rotor kite has no engine at all, and relies on either being carried aloft and dropped from another aircraft, or by being towed into the air behind a car or boat.

## Rotor Configuration

### Number of Blades

A rotary wing is characterised by the number of blades. Typically this is between two and six per driveshaft.

### Number of Rotors

A rotorcraft may have one or more rotors. Various rotor configurations have been used:

- One rotor. Powered rotors require compensation for the torque reaction causing yaw, except in the case of tipjet drive.
- Two rotors. These typically rotate in opposite directions cancelling the torque reaction so that no tail rotor or other yaw stabiliser is needed. These rotors can be laid out as
    - Tandem – One in front of the other.
    - Transverse – Side by side.
    - Coaxial – One rotor disc above the other, with concentric drive shafts.
    - Intermeshing rotors – Twin rotors at an acute angle from each other, whose nearly-vertical driveshafts are geared together to synchronise their rotor blades so that they intermesh, also called a synchropter.
- Three rotors. An uncommon configuration; the 1948 Cierva Air Horse had three rotors as it was not believed a single rotor of sufficient strength could be built for its size. All three rotors turned in the same direction and yaw compensation was provided by inclining each rotor axis to generate rotor thrust components that opposed torque.
- Four rotors. Also referred to as quadcopters/quadrotors, they typically have two rotors turning clockwise and two counter-clockwise.
- More than four rotors. These designs (referred to generally as multirotors, or sometimes individually as hexacopters and octocopters), have matched sets of rotors turning in opposite directions, and uncommon in full-size manned aircraft, but commonly seen in unmanned aerial vehicle systems.

## Jet Aircraft

Jet aircraft are aircraft with jet engines.

Unlike propeller-powered aircraft, jet aircraft normally fly at altitudes as high as 10,000 to 15,000 meters, about 33,000 to 49,000 feet.

At these altitudes, jet engines can achieve maximum efficiency over long distances.

The engines in propeller powered aircraft achieve their maximum efficiency at much lower altitudes.

## Other Jets

Most people use the term 'jet aircraft' to denote gas turbine based airbreathing jet engines, but rockets and scramjets are both also propelled by jet propulsion.

Cruise missiles are single-use unmanned jet aircraft, powered predominately by ramjets or turbojets or sometimes turbofans, but they will often have a rocket propulsion system for initial propulsion.

The fastest airbreathing jet aircraft is the unmanned X-43 scramjet at around Mach 9–10.

The fastest manned (rocket) aircraft is the X-15 at Mach 6.85.

The Space Shuttle, while far faster than the X-43 or X-15, was not regarded as an aircraft during ascent as it was carried ballistically by rocket thrust, rather than the air. During re-entry it was classed (like a glider) as an unpowered aircraft. The first flight was in 1981.

The Bell 533 (1964), Lockheed XH-51 (1965), and Sikorsky S-69 (1977-1981) are examples of compound helicopter designs where jet exhaust added to forward thrust. The Hiller YH-32 Hornet and Fairey Ultra-light Helicopter were among the many helicopters where the rotors were driven by tip jets.

Jet-powered wingsuits exist - powered by model aircraft jet engines - but of short duration and needing to be launched at height.

## Aerodynamics

A McDonnell-Douglas DC-10, an example of a Trijet configuration

Because of the way they work, the typical exhaust speed of jet engines is transonic or faster, therefore most jet aircraft need to fly at high speeds, either supersonic or speeds just below the speed of sound ("transonic") so as to achieve efficient flight. Aerodynamics is therefore an important consideration.

Jet aircraft are usually designed using the Whitcomb area rule, which says that the total area of cross-section of the aircraft at any point along the aircraft from the nose must be approximately the same as that of a Sears-Haack body. A shape with that property minimises the production of shockwaves which would waste energy.

## Jet Engines

Jet engines come in several main types:

- turbojet
- turbofan (which come in two main forms low bypass turbofan and high bypass turbofan)
- rocket

The different types are used for different purposes.

Rockets are the oldest type and are mainly used when extremely high speeds or extremely high altitudes are needed. Due to the extreme, typically hypersonic, exhaust velocity and the necessity of oxidiser being carried on board, they consume propellant extremely quickly. For this reason, they are not practical for routine transportation.

Turbojets are the second oldest type; it has a high, usually supersonic, exhaust speed and low frontal cross-section, and so is best suited to high-speed, usually supersonic, flight. Although once widely used, they are relatively inefficient compared to turboprop and turbofans for subsonic flight. The last major aircraft to use turbojets were Concorde and Tu-144 supersonic transports.

Low bypass turbofans have a lower exhaust speed than turbojets and are mostly used for high sonic, transonic, and low supersonic speeds. High bypass turbofans are used for subsonic aircraft and are quite efficient and are widely used for airliners.

## Flying Characteristics

Jet aircraft fly considerably differently to propeller aircraft.

One difference is that jet engines respond relatively slowly. This complicates takeoff and landing maneuvers. In particular, during takeoff, propeller aircraft engines blow air over their wings and that gives more lift and a shorter takeoff.

## Propulsive Efficiency

In aircraft *overall propulsive efficiency* $\eta$ is the efficiency, in percent, with which the energy contained in a vehicle's propellant is converted into useful energy, to replace losses due to air drag, gravity, and acceleration. It can also be stated as the proportion of the mechanical energy actually used to propel the aircraft. It is always less than 100% because of kinetic energy loss to the exhaust, and less-than-ideal efficiency of the propulsive mechanism, whether a propeller, a jet exhaust, or a fan. In addition, propulsive efficiency is greatly dependent on air density and airspeed.

Mathematically, it is represented as $\eta = \eta_c \eta_p$ where $\eta_c$ is the cycle efficiency and $\eta_p$ is the propulsive efficiency. The cycle efficiency, in percent, is the proportion of energy that can be derived from the energy source that is converted to mechanical energy by the engine.

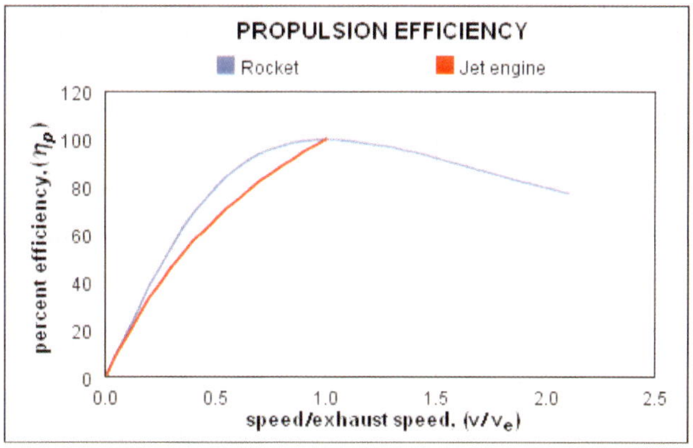


Dependence of the propulsive efficiency ($\eta_p$) upon the vehicle speed/exhaust speed ratio (v/c) for rocket and jet engines.

For jet aircraft the *propulsive efficiency* (essentially energy efficiency) is highest when the engine emits an exhaust jet at a speed that is the same as, or nearly the same as, the vehicle velocity. The exact formula for air-breathing engines as given in the literature, is

$$\eta_p = \frac{2}{1+\frac{c}{v}}$$

where $c$ is the exhaust speed, and $v$ is the speed of the aircraft.

## Range

For a long range jet operating in the stratosphere, the speed of sound is constant, hence flying at fixed angle of attack and constant Mach number causes the aircraft to climb, without changing the value of the local speed of sound. In this case:

$$V = aM$$

where $M$ is the cruise Mach number and $a$ the local speed of sound. The range equation can be shown to be:

$$R = \frac{aM}{c_T}\frac{C_L}{C_D}ln\frac{W_1}{W_2}$$

which is known as the *Breguet range equation* after the French aviation pioneer Louis Charles Breguet.

## Rocket-powered Aircraft

Thrust is the force which moves any aircraft through the air. Thrust is generated by the propulsion system of the aircraft. Different propulsion systems develop thrust in different ways, but all thrust is generated through some application of Newton's third law of motion. For every action there is an equal and opposite reaction. In any propulsion system, a working fluid is accelerated by the system and the reaction to this acceleration produces a force on the system. A general derivation of the thrust equation shows that the amount of thrust generated depends on the mass flow through the engine and the exit velocity of the gas.

During and following World War II, there were a number of rocket- powered aircraft built to explore high speed flight. The X-1A, used to break the "sound barrier", and the X-15 were rocket-powered airplanes. In a rocket engine, fuel and a source of oxygen, called an oxidizer, are mixed and exploded in a combustion chamber. The combustion produces hot exhaust which is passed through a nozzle to accelerate the flow and produce thrust. For a rocket, the accelerated gas, or working fluid, is the hot exhaust produced during combustion. This is a different working fluid than you find in a turbine engine or a propeller powered aircraft. Turbine engines and propellers use air from the atmosphere as the working fluid, but rockets use the combustion exhaust gases. In outer space there is no atmosphere so turbines and propellers can not work there. This explains why a rocket works in space but a turbine engine or a propeller does not work.

There are two main categories of rocket engines; liquid rockets and solid rockets. In a liquid rocket, the propellants, the fuel and the oxidizer, are stored separately as liquids and are pumped into the combustion chamber of the nozzle where burning occurs. In a solid rocket, the propellants are mixed together and packed into a solid cylinder. Under normal temperature conditions, the propellants do not burn; but they will burn when exposed to a source of heat provided by an igniter. Once the burning starts, it proceeds until all the propellant is exhausted. With a liquid rocket, you can stop the thrust by turning off the flow of propellants; but with a solid rocket, you have to destroy the casing to stop the engine. Liquid rockets tend to be heavier and more complex because of the pumps and storage tanks. The propellants are loaded into the rocket just before launch. A solid rocket is much easier to handle and can sit for years before firing.

## Principle of Operation

A rocket carries all the fuel and oxidizing agent required to operate the engine and hence the engine may operate both in air and in a vacuum. The rocket engine exerts a forward thrust ae a reaction to Expelling the exhaust products rearward at high velocity as a jet. This thrust can be expressed in terms of effective exhaust jet velocity ~ fuel consumption by equating thrust to rate of change of the momentum of the exhaust gases being expended. The variation of fuel consumption with effective jet velocity. It is apparent that high jet velocities are required for low specific fuel consumption.

## Rocket Velocity Limitation

The final velocity of a rocket-powered aircraft is dependent upon air resistance, effective jet velocity and percentage of' fuel load. Even in the absence of air resistance a limit is set by the fuel load and the effective jet velocity on the final speed that may be obtained. The maximum velocity attained by a rocket –powered aircraft in an empty gravity-free space may be determined by equating the thrust to the product of instantaneous rocket mass and instantaneous rocket acceleration. This limiting velocity as a function of' both fuel load and jet velocity. and. shows that the speed. limitationis directly proportional to jet velocity and is critically dependent on fuel load. For instance, at a 90-pereent fuel load the value of limiting velocity is 3·5 times as large as the value at a 45-percent fuel~ 1oad.

## An Example

The X-15 was a rocket-powered aircraft 50 ft long with a wingspan of 22 ft. It was a missile-shaped vehicle with an unusual wedge-shaped vertical tail, thin stubby wings, and unique side fairings that extended along the side of the fuselage. The X-15 weighed about 14,000 lb empty and approximately 34,000 lb at launch. The XLR-99 rocket engine, manufactured by Thiokol Chemical Corp., was pilot controlled and was capable of developing 57,000 lb of thrust. North American Aviation made three X-15 aircraft for the program.

The X-15 research aircraft was developed to provide in-flight information and data on aerodynamics, structures, flight controls, and the physiological aspects of high-speed, high-altitude flight. A follow on program used the aircraft as a testbed to carry various scientific experiments beyond the Earth's atmosphere on a repeated basis.

For flight in the dense air of the usable atmosphere, the X-15 used conventional aerodynamic controls such as rudder surfaces on the vertical stabilizers to control yaw and canted horizontal surfaces on the tail to control pitch when moving in synchronization or roll when moved differentially.

For flight in the thin air outside of the appreciable Earth's atmosphere, the X-15 used a reaction control system. Hydrogen peroxide thrust rockets located on the nose of the aircraft provided pitch and yaw control. Those on the wings provided roll control.

Because of the large fuel consumption, the X-15 was air launched from a B-52 aircraft at 45,000 ft and a speed of about 500 mph. Depending on the mission, the rocket engine provided thrust for the first 80 to 120 sec of flight. The remainder of the normal 10 to 11 min. flight was powerless and ended with a 200-mph glide landing.

Generally, one of two types of X-15 flight profiles was used; a high-altitude flight plan that called for the pilot to maintain a steep rate of climb, or a speed profile that called for the pilot to push over and maintain a level altitude.

The X-15 was flown over a period of nearly 10 years - June 1959 to Oct. 1968 - and set the world's unofficial speed and altitude records of 4,520 mph (Mach 6.7) and 354,200 ft in a program to investigate all aspects of manned hypersonic flight. Information gained from the highly successful X-15 program contributed to the development of the Mercury, Gemini, and Apollo manned spaceflight programs, and also the Space Shuttle program.

The X-15s made a total of 199 flights, and were manufactured by North American Aviation.

# Military Aircraft

Military aircraft, any type of aircraft that has been adapted for military use.

Aircraft have been a fundamental part of military power since the mid-20th century. Generally speaking, all military aircraft fall into one of the following categories: fighters, which secure control of essential airspaces by driving off or destroying enemy aircraft; bombers, which are larger, heavier, and less-maneuverable craft designed to attack surface targets with bombs or missiles; ground-support, or attack, aircraft, which operate at lower altitudes than bombers and air-superiority fighters and attack tanks, troop formations, and other ground targets; transport and cargo planes, big-bodied craft with large amounts of interior space for carrying weapons, equipment, supplies, and troops over moderate or long distances; helicopters, which are rotary-winged aircraft used for ground support, for transporting assault troops, and for short-distance transport and surveillance; and unmanned aerial vehicles, which are remotely controlled or autonomously guided aircraft that carry sensors, target designators, electronic transmitters, and even offensive weapons.

## Combat Aircraft

Combat aircraft, or "Warplanes", are divided broadly into *multi-role, fighters, bombers, attackers*, and *electronic warfare support*.

Variations exist between them, including fighter-bombers, such as the MiG-23 ground-attack aircraft and the Soviet Ilyushin Il-2 Sturmovik. Also included among combat aircraft are long-range maritime patrol aircraft, such as the Hawker Siddeley Nimrod and the S-3 Viking that are often equipped to attack with anti-ship missiles and anti-submarine weapons.

## Fighter Aircraft

The primary role of fighters is destroying enemy aircraft in air-to-air combat, as part of both offensive and defensive counter air operations. Many fighters also possess a degree of ground attack capability, allowing them to perform surface attack and close air support missions. In addition to their counter air duties they are tasked to perform escort mission for bombers or other aircraft. Fighters are capable of carrying a variety of weapons, including machine guns, cannons, rockets, guided missiles, and bombs. Many modern fighters can attack enemy fighters from a great distance, before the enemy even sees or detects them. Examples of fighters include the F-22 Raptor, F-15 Eagle, and Su-27.

A USAF F-22A Raptor firing an AIM-120 air-to-air missile

## Bomber Aircraft

Bombers are normally larger, heavier, and less maneuverable than fighter aircraft. They are capable of carrying large payloads of bombs, torpedoes or cruise missiles. Bombers are used almost exclusively for ground attacks and not fast or agile enough to take on enemy fighters head-to-head. A few have a single engine and require one pilot to operate and others have two or more engines and require crews of two or more. A limited number of bombers, such as the B-2 Spirit, have stealth capabilities that keep them from being detected by enemy radar. An example of a conventional modern bomber would be the B-52 Stratofortress. An example of a World War II bomber would be a B-17 Flying Fortress. Bombers include light bombers, medium bombers, heavy bombers, dive bombers, and torpedo bombers.

A USAF B-2A Spirit

## Attack Aircraft

A USAF A-10 Thunderbolt II firing an AGM-65

Attack aircraft can be used to provide support for friendly ground troops. Some are able to carry conventional or nuclear weapons far behind enemy lines to strike priority ground targets. Attack helicopters attack enemy armor and provide close air support for ground troops. An example historical ground-attack aircraft is the Soviet Ilyushin Il-2 Shturmovik. Several types of transport airplanes have been armed with sideways firing weapons as gunships for ground attack. These include the AC-47 and AC-130 aircraft.

## Electronic Warfare Aircraft

A USAF EF-111A Raven

An electronic warfare aircraft is a military aircraft equipped for electronic warfare (EW) - i.e. degrading the effectiveness of enemy radar and radio systems. They are generally modified versions of other pre-existing aircraft. A recent example would be the Boeing EA-18G Growler, which is a modified version of the Boeing F/A-18F Super Hornet.

## Maritime Patrol Aircraft

A maritime patrol aircraft fixed-wing military aircraft designed to operate for long durations over water in maritime patrol roles—in particular anti-submarine, anti-ship and search and rescue. Some patrol aircraft were designed for this purpose, like the Kawasaki P-1. Many others are modified designs of pre-existing aircraft, such as the Boeing P-8 Poseidon, which is based on the Boeing 737-800 airliner.

A USN P-8 Poseidon

## Multirole Combat Aircraft

Many combat aircraft today have a multirole ability. Normally only applying to fixed-wing aircraft, this term signifies that the plane in question can be a fighter or a bomber, depending on what the mission calls for. An example of a multirole design is the F-15E Strike Eagle, F/A-18 Hornet, F-35 Lightning II. A World War II example would be the P-38 Lightning.

A USAF F-15E Strike Eagle dropping a PGB

## Non-combat Aircraft

Non-combat roles of military aircraft include search and rescue, reconnaissance, observation/surveillance, Airborne Early Warning and Control, transport, training, and aerial refueling.

Many civil aircraft, both fixed wing and rotary wing, have been produced in separate models for military use, such as the civilian Douglas DC-3 airliner, which became the military C-47 Skytrain, and British "Dakota" transport planes, and decades later, the USAF's AC-47 aerial gunships. Even the fabric-covered two-seat Piper J3 Cub had a military version. Gliders and balloons have also been used as military aircraft; for example, balloons were used for observation during the American Civil War and during World War I, and military gliders were used during World War II to deliver ground troops in airborne assaults.

## Military Transport Aircraft

Military transport (logistics) aircraft are primarily used to transport troops and war supplies. Cargo can be attached to pallets, which are easily loaded, secured for flight, and quickly unloaded for delivery. Cargo also may be discharged from flying aircraft on parachutes, eliminating the need for landing. Also included in this category are aerial tankers; these planes can refuel other aircraft while in flight. An example of a transport aircraft is the C-17 Globemaster III. A World War II example would be the C-47. An example of a tanker craft would be the KC-135 Stratotanker. Helicopters and gliders can transport troops and supplies to areas where other aircraft would be unable to land.

A USAF C-17A Globemaster III military transport aircraft

Calling a military aircraft a "cargo plane" is incorrect, because military *transport planes* also carry paratroopers and other soldiers.

## Airborne Early Warning and Control

A USAF E-3 Sentry

An airborne early warning and control (AEW&C) system is an airborne radar system designed to detect aircraft, ships and ground vehicles at long ranges and control and command the battle space in an air engagement by directing fighter and attack aircraft strikes. AEW&C units are also used to carry out surveillance, including over ground targets and frequently perform C2BM (command and control, battle management) functions similar to an Airport Traffic Controller given military command over other forces. Used at a high altitude, the radars on the aircraft allow the operators to distinguish between friendly and hostile aircraft hundreds of miles away.

An Indian Air Force Beriev A-50EI Mainstay

AEW&C aircraft are used for both defensive and offensive air operations, and are to the NATO and USA forces trained or integrated Air Forces what the Command Information Center is to a Navy Warship, plus a highly mobile and powerful radar platform. The system is used offensively to direct fighters to their target locations, and defensively in order to counterattacks by enemy forces, both air and ground. So useful is the advantage of command and control from a high altitude, the United States Navy operates AEW&C aircraft off its Supercarriers to augment and protect its carrier Command Information Centers (CICs).

AEW&C is also known by the older terms "airborne early warning" (AEW) and "airborne warning and control system" (AWACS, /ˈeɪwæks/ ay-waks) although AWACS is the name of a specific system currently used by NATO and the USAF and is often used in error to describe similar systems.

### Reconnaissance and Surveillance Aircraft

A RAF Raytheon Sentinel with a radar pod

Reconnaissance aircraft are primarily used to gather intelligence. They are equipped with cameras and other sensors. These aircraft may be specially designed or may be modified from a basic fighter or bomber type. This role is increasingly being filled by satellites and unmanned aerial vehicles (UAVs).

Surveillance and observation aircraft use radar and other sensors for battlefield surveillance, airspace surveillance, maritime patrol and artillery spotting. They include modified civil aircraft designs, moored balloons and UAVs.

### Experimental Aircraft

Experimental aircraft are designed in order to test advanced aerodynamic, structural, avionic, or propulsion concepts. These are usually well instrumented, with performance data telemetered on radio-frequency data links to ground stations located at the test ranges where they are flown. An example of an experimental aircraft is the Bristol 188.

## Supersonic Aircraft

Supersonic flight is one of the four speeds of flight. They are called the regimes of flight. The regimes of flight are subsonic, transonic, supersonic and hypersonic.

Vehicles that fly at supersonic speeds are flying faster than the speed of sound. The speed of sound is about 768 miles per hour (1,236 kilometers per hour) at sea level. These speeds are referred to by Mach numbers. The Mach number is the ratio of the speed of the aircraft to the speed of sound. Flight that is faster than Mach 1 is supersonic. Supersonic includes speeds up to five times faster than the speed of sound, or Mach 5.

In 1947, Air Force Capt. Charles E. "Chuck" Yeager became the first person to fly an aircraft faster than the speed of sound.

## Design Principles

Supersonic flight brings with it substantial technical challenges, as the aerodynamics of supersonic flight are dramatically different from those of subsonic flight (i.e., flight at speeds slower than that of sound). In particular, aerodynamic drag rises sharply as the aircraft passes the transonic regime, requiring much greater engine power and more streamlined airframes.

## Wings

To keep drag low, wing span must be limited, which also reduces the aerodynamic efficiency when flying slowly. Since a supersonic aircraft must take off and land at a relatively slow speed, its aerodynamic design must be a compromise between the requirements for both ends of the speed range.

One approach to resolving this compromise is the use of a variable-geometry wing, commonly known as the "swing-wing," which spreads wide for low-speed flight and then sweeps sharply, usually backwards, for supersonic flight. However, swinging affects the longitudinal trim of the aircraft and the swinging mechanism adds weight and cost.

## Heating

Another problem is the heat generated by friction as the air flows over the aircraft. Most supersonic designs use aluminium alloys such as Duralumin, which are cheap and easy to work but lose their strength quickly at high temperatures. This limits maximum speed to around Mach 2.2.

Most supersonic aircraft, including many military fighter aircraft, are designed to spend most of their flight at subsonic speeds, and only to exceed the speed of sound for short periods such as when intercepting an enemy aircraft. A smaller number, such as the Lockheed SR-71 Blackbird reconnaissance aircraft and the Concorde supersonic airliner, have been designed to cruise continuously at speeds above the speed of sound, and with these designs the problems of supersonic flight are more severe.

## Engines

Some early supersonic aircraft, including the first, relied on rocket power to provide the necessary thrust, although rockets burn a lot of fuel and so flight times were short. Early turbojets were more fuel-efficient but did not have enough thrust and some experimental aircraft were fitted with both a turbojet for low-speed flight and a rocket engine for supersonic flight. The invention of the afterburner, in which extra fuel is burned in the jet exhaust, made these mixed powerplant types obsolete. The turbofan engine passes additional cold air around the engine core, further increasing its fuel efficiency, and supersonic aircraft today are powered by turbofans fitted with after burners.

Supersonic aircraft usually use low bypass turbofans as they have acceptable efficiency below the speed of sound as well as above; or if supercruise is needed turbojet engines may be desirable as they give less nacelle drag at supersonic speeds. The Pratt & Whitney J58 engines of the Lockheed

SR-71 Blackbird operated in 2 ways, taking off and landing as turbojets with no bypass, but bypassing some of the compressor air to the afterburner at higher speeds. This allowed the Blackbird to fly at over Mach 3, faster than any other production aircraft. The heating effect of air friction at these speeds meant that a special fuel had to be developed which did not break down in the heat and clog the fuel pipes on its way to the burner.

Another high-speed powerplant is the ramjet. This needs to be flying fairly fast before it will work at all.

## Transonic Flight

Airflow can speed up or slow down locally at different points over an aircraft. In the region around Mach 1, some areas may experience supersonic flow while others are subsonic. This regime is called transonic flight. As the aircraft speed changes, pressure waves will form or move around. This can affect the trim, stability and controllability of the aircraft, and the designer needs to ensure that these effects are taken into account at all speeds.

## Hypersonic Flight

Flight at speeds above about Mach 5 is often referred to as hypersonic. In this region the problems of drag and heating are even more acute. It is difficult to make materials which can stand the forces and temperatures generated by air resistance at these speeds, and hypersonic flight for any significant length of time has not yet been achieved.

## Sonic Boom

A sonic boom is the sound associated with the shock waves created whenever an object traveling through the air travels faster than the speed of sound. Sonic booms generate significant amounts of sound energy, sounding similar to an explosion or a thunderclap to the human ear. The crack of a supersonic bullet passing overhead or the crack of a bullwhip are examples of a sonic boom in miniature.

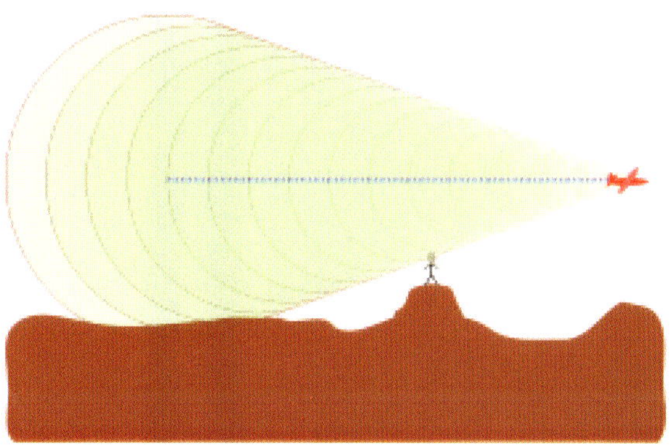

A sonic boom produced by an aircraft moving at M=2.92, calculated from the cone angle of 20 degrees. An observer hears nothing until the shock wave, on the edges of the cone, crosses their location.

Sonic booms due to large supersonic aircraft can be particularly loud and startling, tend to awaken people, and may cause minor damage to some structures. They led to prohibition of routine supersonic flight over land. Although they cannot be completely prevented, research suggests that with careful shaping of the vehicle the nuisance due to them may be reduced to the point that overland supersonic flight may become a practical option.

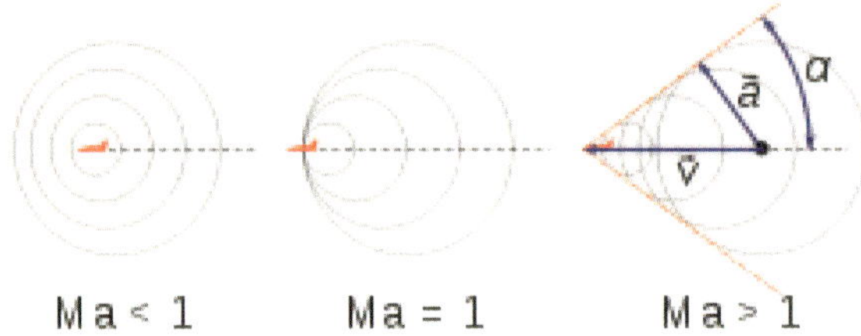


Mach cone angle

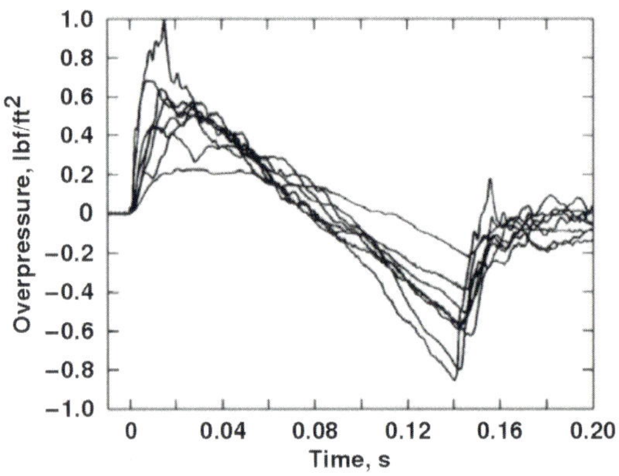


NASA data showing N-wave signature.

## Supersonic Transports

The fuselage of Concorde had an extremely high fineness ratio.

A supersonic transport (SST) is a civil aircraft designed to transport passengers at speeds greater than the speed of sound. The only supersonic civilian aircraft to see service were the Soviet produced Tupolev Tu-144 which first flew in 1968 and was retired in 1997; and the Franco-British produced Concorde, which first flew in 1969 and remained in service until 2003. Since 2003, there have been no supersonic civilian aircraft in service.

A key feature of these designs is the ability to maintain supersonic cruise for long periods, so low drag is essential to limit fuel consumption to a practical and economic level. As a consequence, these airframes are highly streamlined and the wings have a very short span. The requirement for low speeds when taking off and landing is met by using vortex lift: as the aircraft slows, lift must be restored by raising the nose to increase the angle of attack of the wing. The sharply swept leading edge causes the air to twist as it flows over the wing, speeding up the airflow locally and maintaining lift.

Other SST projects have included:

- British – Bristol Type 223
- French – Sud Aviation Super-Caravelle
- US – Convair Model 58-9, Boeing 2707, Lockheed L-2000, Douglas 2229, SAI Quiet Supersonic Transport, High Speed Civil Transport
- USSR – Tupolev Tu-244, Tupolev Tu-444
- Russian-American – Sukhoi-Gulfstream S-21

## Supersonic Business jet

Aerion SBJ model

Supersonic business jets (SSBJ) are a proposed class of small supersonic aircraft. None have yet flown.

Typically intended to transport about ten passengers, SSBJs are about the same size as traditional subsonic business jets.

Projects for both large-scale and business jet (see lower) passenger supersonic and hypersonic airliners (Aerion SBJ, Spike S-512, HyperMach SonicStar, Next Generation Supersonic Transport, Tupolev Tu-444, Gulfstream X-54, LAPCAT, Reaction Engines A2, Zero Emission Hyper Sonic Transport, SpaceLiner, etc.) were proposed and now are under development.

## Supersonic Strategic Bombers

XB-70 Valkyrie

Tupolev Tu-22M3

Tupolev Tu-160

B-1B Lancer

Convair B-58A Hustler

A strategic bomber must carry a large bomb load over long distances. Consequently, it is a large aircraft typically with an empty weight exceeding 25,000 kg. Some have also been designed for related roles such as strategic reconnaissance and anti-shipping strike.

Typically the aircraft will cruise subsonically for most of its flight to conserve fuel, before accelerating to supersonic speed for its bombing run.

Few supersonic strategic bombers have entered service. The earliest type, the Convair B-58 Hustler, first flew in 1956 and the most recent, the Rockwell B-1B Lancer, in 1983. Although this and a few other types are still in service today, none remains in production.

## Supersonic Strategic Reconnaissance

Some supersonic strategic bombers, such as the Sukhoi T-4 are also capable of the reconnaissance role (although the Sukhoi remained a prototype).

The Lockheed SR-71 Blackbird was specifically designed for the role, and was a larger development of the Lockheed A-12 attack aircraft which first flew in 1962.

## Supersonic Fighter/Attack Jets

Supersonic fighters and related aircraft are sometimes called fast jets. They make up the overwhelming majority of supersonic aircraft and some, such as the Mikoyan-Gurevich MiG-21, Lockheed F-104 Starfighter and Dassault Mirage III, have been produced in large numbers.

Many military supersonic fighters and similar aircraft of fourth- and fifth- generations are under development in several countries, including Russia, China, Japan, South Korea, India, Iran and the United States.

## Experimental Aircraft

Amateur-built aircraft are built by individuals and licensed by the Federal Aviation Administration (FAA) as "Experimental." The Experimental designation has been in existence for more than five decades. It defines aircraft that are used for non-commercial, recreational purposes such as education or personal use.

Under FAA regulations, if an individual builds at least 51 percent of an aircraft, the aircraft is eligible to be registered in the amateur-built category. They are available in kits (where some of the airplane is already fabricated), or plans (where the builder purchases or manufactures all the parts and assembles them). These airplanes are also commonly known as "homebuilts," for the obvious reason that many individuals construct these aircraft at home, often in their garages.

Currently, more than 33,000 amateur-built/homebuilt aircraft are licensed by FAA. They represent proven aircraft designs that have been flown safely for many years.

### Construction of Amateur-built/homebuilt Aircraft

People from all walks of life, including astronauts, airline pilots, military jet pilots, machinists, welders, professional people and others.

### Purpose for Building

A variety of reasons: a personal challenge, education, performance, or to invest "sweat equity" into the cost of an airplane.

Costs range from under $10,000 to more than $100,000 based on desired performance characteristics and optional engine and avionics packages. By comparison, a new factory-built Cessna 172 costs more than $250,000.

Many amateur-built/homebuilt aircraft utilize composite materials that help create airplanes that are lighter, faster and more fuel efficient than similar production aircraft.

### Time Taken

An average amateur-built/homebuilt aircraft will take between 1,000 and 3,000 hours to complete. Some individuals build their airplane in less than a year; others may take a decade or more.

### Aircraft Regulation

All amateur-built/homebuilt airplanes must be registered with the Federal Aviation Administration (FAA). These airplanes must be inspected by an FAA inspector or a designated inspector before an airworthiness certificate can be issued.

This is a fairly rigorous process. The builder(s) must provide logs of when, where and how construction took place, along with supporting documents and photographs. If the aircraft passes this inspection, a pilot must fly between 25-40 hours of test flights in specific non-populated areas to make sure all components are operating properly. Only after that test time is flown may passengers be flown in the aircraft.

In addition, an amateur-built airplane is subject to condition inspections every 12 months, the same scrutiny required of small production aircraft.

## Pilot License

Yes. Pilots of amateur-built/homebuilt aircraft must earn and maintain the same federal pilot training and ratings as those who fly factory-built aircraft such as Cessnas, Pipers, and Beechcrafts. They also must follow all appropriate federal regulations during each of their flights.

## Meaning of "Experimental"

The term "Experimental" is actually a bit of a misnomer; it refers to the FAA category in which the airplane is registered, not the exclusivity or the use of the airplane.

While there are a handful of homebuilt aircraft that are original designs, the vast majority of homebuilt airplanes are built using standardized, tried- and-true kits or plans that have been successfully constructed thousands of times.

When the current homebuilt aircraft rules were first introduced in the early 1950s, there was difficulty finding a category where the finished aircraft could be registered. After all, the airplanes were not factory-built, such as Cessnas or Pipers, nor were they transport aircraft (airliners) or military aircraft. Federal officials saw the most practical category as Experimental, and created a new subcategory called "amateur-built."

FAA's Experimental category also includes nearly 10 other subcategories, including aircraft used for crew training, air racing, and historic aircraft (such as World War II military aircraft) flown to air shows and exhibitions.

## Safety

Studies by FAA and the National Transportation Safety Board (NTSB) show that amateur-built/homebuilt aircraft have an accident rate less than one percentage point higher than the general aviation fleet. In fact, the accident rate for amateur-built/homebuilt aircraft is dropping. The total number of registered homebuilt aircraft has doubled since 1994, and the total hours flown have increased by 123 percent, while the total number of accidents has stayed virtually the same.

Another good barometer of safety is insurance rates. Companies that insure both homebuilts and production aircraft charge about the same rates for owners of either type of airplane. That indicates a similar level of risk.

## References


- Hamilton-Paterson, James (2010). Empire of the Clouds: When Britain's Aircraft Ruled the World (1St edition reprint ed.). Faber & Faber. p. 304. ISBN 0571247946
- Cessna rolls out first production unit of new Citation X,"Archived 22 November 2016 at the Wayback Machine. Apr 15, 2013, Wichita Business Journal, retrieved November 22, 2016
- Crane, Dale: Dictionary of Aeronautical Terms, third edition, page 194. Aviation Supplies & Academics, 1997. ISBN 1-56027-287-2
- [ "Airlander 10, the world's largest aircraft, takes off for the first time,"] August 19, 2016, CBS News (TV) retrieved November 22, 2016
- Aviation Publishers Co. Limited, From the Ground Up, page 10 (27th revised edition) ISBN 0-9690054-9-0
- World's Second Largest Aircraft," Archived 22 November 2016 at the Wayback Machine. July 28th, 2013, NASA, retrieved November 22, 2016
- Hecht, Heinrich. The World's First Turbojet Fighter – Messerschmitt Me 262. Atglen, Pennsylvania: Schiffer Publishing, 1990. ISBN 0-88740-234-8

2

# Understanding Aerodynamics

Aerodynamics is the study of the motion of air around a soild object. This chapter has been carefully written to provide an easy understanding of the fundamentals of aerodynamics by elucidating topics such as compressible flow, center of pressure, boundary layer, supersonic speed and hypersonic speed, besides others.

Aerodynamics is the study of how gases interact with moving bodies. Because the gas that we encounter most is air, aerodynamics is primarily concerned with the forces of drag and lift, which are caused by air passing over and around solid bodies. Engineers apply the principles of aerodynamics to the designs of many different things, including buildings, bridges and even soccer balls; however, of primary concern is the aerodynamics of aircraft and automobiles.

Aerodynamics comes into play in the study of flight and the science of building and operating an aircraft, which is called aeronautics. Aeronautical engineers use the fundamentals of aerodynamics to design aircraft that fly through the Earth's atmosphere.

## Aerodynamic Drag

The most significant aerodynamic force that applies to nearly everything that moves through the air is drag. Drag is the force that opposes an aircraft's motion through the air, according to NASA. Drag is generated in the direction the air is moving when it encounters a solid object. In most cases, such as in automobiles and aircraft, drag is undesirable because it takes power to overcome it. There are, however, some cases when drag is beneficial, such as with parachutes, for example.

To describe the amount of drag on an object, we use a value called the drag coefficient ($c_d$). This number depends not only on the shape of the object but also on other factors, such as its speed and surface roughness, the density of the air and whether the flow is laminar (smooth) or turbulent. Forces that affect drag include the air pressure against the face of the object, the friction along the sides of the object and the relatively negative pressure, or suction, on the back of the object. For example, $c_d$ for a flat plate moving face-on through the air is about 1.3, a face-on cube is about 1, a sphere is about 0.5 and a teardrop shape is about 0.05. The drag coefficient for modern automobiles is 0.25 to 0.35, and for aircraft it is 0.01 to 0.03. Calculating $c_d$ can be complicated. For this reason, it is usually determined by computer simulations or wind tunnel experiments.

## Fundamental Concepts

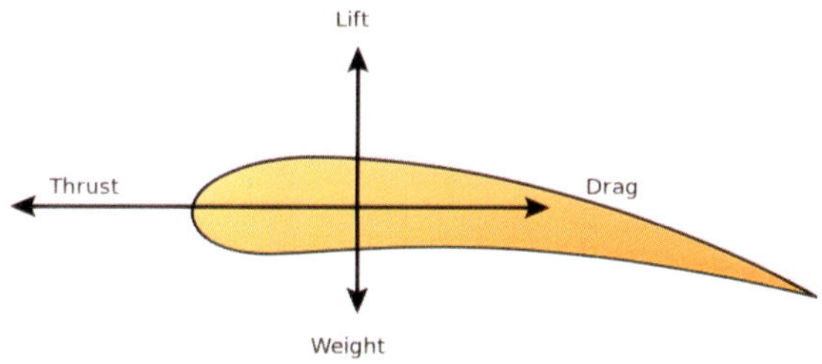


Forces of flight on an airfoil

Understanding the motion of air around an object (often called a flow field) enables the calculation of forces and moments acting on the object. In many aerodynamics problems, the forces of interest are the fundamental forces of flight: lift, drag, thrust, and weight. Of these, lift and drag are aerodynamic forces, i.e. forces due to air flow over a solid body. Calculation of these quantities is often founded upon the assumption that the flow field behaves as a continuum. Continuum flow fields are characterized by properties such as flow velocity, pressure, density, and temperature, which may be functions of position and time. These properties may be directly or indirectly measured in aerodynamics experiments or calculated starting with the equations for conservation of mass, momentum, and energy in air flows. Density, flow velocity, and an additional property, viscosity, are used to classify flow fields.

## Flow Classification

Flow velocity is used to classify flows according to speed regime. Subsonic flows are flow fields in which the air speed field is always below the local speed of sound. Transonic flows include both regions of subsonic flow and regions in which the local flow speed is greater than the local speed of sound. Supersonic flows are defined to be flows in which the flow speed is greater than the speed of sound everywhere. A fourth classification, hypersonic flow, refers to flows where the flow speed is much greater than the speed of sound. Aerodynamicists disagree on the precise definition of hypersonic flow.

Compressible flow accounts for varying density within the flow. Subsonic flows are often idealized as incompressible, i.e. the density is assumed to be constant. Transonic and supersonic flows are compressible, and calculations that neglect the changes of density in these flow fields will yield inaccurate results.

Viscosity is associated with the frictional forces in a flow. In some flow fields, viscous effects are very small, and approximate solutions may safely neglect viscous effects. These approximations are called inviscid flows. Flows for which viscosity is not neglected are called viscous flows. Finally, aerodynamic problems may also be classified by the flow environment. External aerodynamics is the study of flow around solid objects of various shapes (e.g. around an airplane wing), while internal aerodynamics is the study of flow through passages inside solid objects (e.g. through a jet engine).

## Continuum Assumption

Unlike liquids and solids, gases are composed of discrete molecules which occupy only a small fraction of the volume filled by the gas. On a molecular level, flow fields are made up of the collisions of many individual of gas molecules between themselves and with solid surfaces. However, in most aerodynamics applications, the discrete molecular nature of gases is ignored, and the flow field is assumed to behave as a continuum. This assumption allows fluid properties such as density and flow velocity to be defined everywhere within the flow.

The validity of the continuum assumption is dependent on the density of the gas and the application in question. For the continuum assumption to be valid, the mean free path length must be much smaller than the length scale of the application in question. For example, many aerodynamics applications deal with aircraft flying in atmospheric conditions, where the mean free path length is on the order of micrometers and where the body is orders of magnitude larger. In these cases, the length scale of the aircraft ranges from a few meters to a few tens of meters, which is much larger than the mean free path length. For such applications, the continuum assumption is reasonable. The continuum assumption is less valid for extremely low-density flows, such as those encountered by vehicles at very high altitudes (e.g. 300,000 ft/90 km) or satellites in Low Earth orbit. In those cases, statistical mechanics is a more accurate method of solving the problem than is continuum aerodynamics. The Knudsen number can be used to guide the choice between statistical mechanics and the continuous formulation of aerodynamics.

## Conservation Laws

The assumption of a fluid continuum allows problems in aerodynamics to be solved using fluid dynamics conservation laws. Three conservation principles are used:

1. Conservation of mass: In fluid dynamics, the mathematical formulation of this principle is known as the mass continuity equation, which requires that mass is neither created nor destroyed within a flow of interest.

2. Conservation of momentum: In fluid dynamics, the mathematical formulation of this principle can be considered an application of Newton's Second Law. Momentum within a flow is only changed by the work performed on the system by external forces, which may include both surface forces, such as viscous (frictional) forces, and body forces, such as weight. The momentum conservation principle may be expressed as either a vector equation or separated into a set of three scalar equations (x,y,z components). In its most complete form, the momentum conservation equations are known as the Navier-Stokes equations. The Navier-Stokes equations have no known analytical solution and are solved in modern aerodynamics using computational techniques. Because of the computational cost of solving these complex equations, simplified expressions of momentum conservation may be appropriate for specific applications. The Euler equations are a set of momentum conservation equations which neglect viscous forces and may be used in cases where the effect of viscous forces is expected to be small. Additionally, Bernoulli's equation is a solution to the momentum conservation equation of an inviscid flow that neglects gravity.

3. Conservation of energy: The energy conservation equation states that energy is neither created nor destroyed within a flow, and that any addition or subtraction of energy to a volume in the flow is caused by the fluid flow, by heat transfer, or by work into and out of the region of interest.

The ideal gas law or another such equation of state is often used in conjunction with these equations to form a determined system that allows the solution for the unknown variables.

## Branches of Aerodynamics

Aerodynamic problems are classified by the flow environment or properties of the flow, including flow speed, compressibility, and viscosity. *External* aerodynamics is the study of flow around solid objects of various shapes. Evaluating the lift and drag on an airplane or the shock waves that form in front of the nose of a rocket are examples of external aerodynamics. *Internal* aerodynamics is the study of flow through passages in solid objects. For instance, internal aerodynamics encompasses the study of the airflow through a jet engine or through an air conditioning pipe.

Aerodynamic problems can also be classified according to whether the flow speed is below, near or above the speed of sound. A problem is called subsonic if all the speeds in the problem are less than the speed of sound, transonic if speeds both below and above the speed of sound are present (normally when the characteristic speed is approximately the speed of sound), supersonic when the characteristic flow speed is greater than the speed of sound, and hypersonic when the flow speed is much greater than the speed of sound. Aerodynamicists disagree over the precise definition of hypersonic flow; a rough definition considers flows with Mach numbers above 5 to be hypersonic.

The influence of viscosity on the flow dictates a third classification. Some problems may encounter only very small viscous effects, in which case viscosity can be considered to be negligible. The approximations to these problems are called inviscid flows. Flows for which viscosity cannot be neglected are called viscous flows.

## Incompressible Aerodynamics

An incompressible flow is a flow in which density is constant in both time and space. Although all real fluids are compressible, a flow is often approximated as incompressible if the effect of the density changes cause only small changes to the calculated results. This is more likely to be true when the flow speeds are significantly lower than the speed of sound. Effects of compressibility are more significant at speeds close to or above the speed of sound. The Mach number is used to evaluate whether the incompressibility can be assumed, otherwise the effects of compressibility must be included.

## Subsonic Flow

Subsonic (or low-speed) aerodynamics describes fluid motion in flows which are much lower than the speed of sound everywhere in the flow. There are several branches of subsonic flow but one special case arises when the flow is inviscid, incompressible and irrotational. This case is called potential flow and allows the differential equations that describe the flow to be a simplified version of the equations of fluid dynamics, thus making available to the aerodynamicist a range of quick and easy solutions.

In solving a subsonic problem, one decision to be made by the aerodynamicist is whether to incorporate the effects of compressibility. Compressibility is a description of the amount of change of density in the flow. When the effects of compressibility on the solution are small, the assumption that density is constant may be made. The problem is then an incompressible low-speed aerodynamics problem. When the density is allowed to vary, the flow is called compressible. In air, compressibility effects are usually ignored when the Mach number in the flow does not exceed 0.3 (about 335 feet (102 m) per second or 228 miles (366 km) per hour at 60 °F (16 °C)). Above Mach 0.3, the problem flow should be described using compressible aerodynamics.

## Compressible Aerodynamics

According to the theory of aerodynamics, a flow is considered to be compressible if the density changes along a streamline. This means that – unlike incompressible flow – changes in density are considered. In general, this is the case where the Mach number in part or all of the flow exceeds 0.3. The Mach 0.3 value is rather arbitrary, but it is used because gas flows with a Mach number below that value demonstrate changes in density of less than 5%. Furthermore, that maximum 5% density change occurs at the stagnation point (the point on the object where flow speed is zero), while the density changes around the rest of the object will be significantly lower. Transonic, supersonic, and hypersonic flows are all compressible flows.

## Transonic Flow

The term Transonic refers to a range of flow velocities just below and above the local speed of sound (generally taken as Mach 0.8–1.2). It is defined as the range of speeds between the critical Mach number, when some parts of the airflow over an aircraft become supersonic, and a higher speed, typically near Mach 1.2, when all of the airflow is supersonic. Between these speeds, some of the airflow is supersonic, while some of the airflow is not supersonic.

## Supersonic Flow

Supersonic aerodynamic problems are those involving flow speeds greater than the speed of sound. Calculating the lift on the Concorde during cruise can be an example of a supersonic aerodynamic problem.

Supersonic flow behaves very differently from subsonic flow. Fluids react to differences in pressure; pressure changes are how a fluid is "told" to respond to its environment. Therefore, since sound is in fact an infinitesimal pressure difference propagating through a fluid, the speed of sound in that fluid can be considered the fastest speed that "information" can travel in the flow. This difference most obviously manifests itself in the case of a fluid striking an object. In front of that object, the fluid builds up a stagnation pressure as impact with the object brings the moving fluid to rest. In fluid traveling at subsonic speed, this pressure disturbance can propagate upstream, changing the flow pattern ahead of the object and giving the impression that the fluid "knows" the object is there by seemingly adjusting its movement and is flowing around it. In a supersonic flow however, the pressure disturbance cannot propagate upstream. Thus, when the fluid finally reaches the object it strikes it and the fluid is forced to change its properties – temperature, density, pressure, and Mach number—in an extremely violent and irreversible fashion called a shock wave. The presence of shock waves, along with the compressibility effects of high-flow velocity fluids, is the central difference between the supersonic and subsonic aerodynamics regimes.

### Hypersonic Flow

In aerodynamics, hypersonic speeds are speeds that are highly supersonic. In the 1970s, the term generally came to refer to speeds of Mach 5 (5 times the speed of sound) and above. The hypersonic regime is a subset of the supersonic regime. Hypersonic flow is characterized by high temperature flow behind a shock wave, viscous interaction, and chemical dissociation of gas.

### Associated Terminology

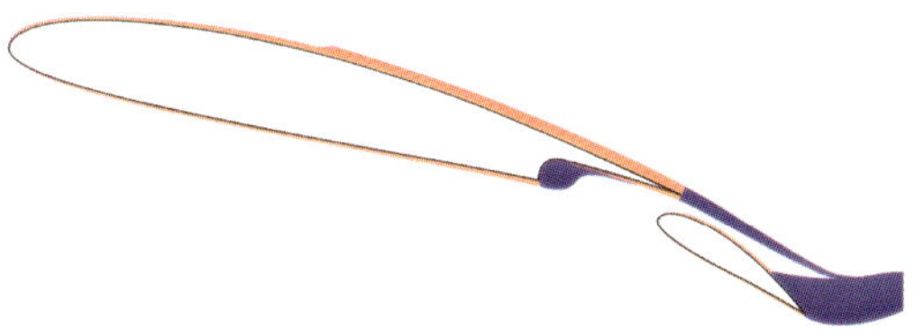

Different types flow analysis around an airfoil:

▯ Potential flow theory ▮ Boundary layer flow theory ▮ Turbulent wake analysis

The incompressible and compressible flow regimes produce many associated phenomena, such as boundary layers and turbulence.

### Boundary Layers

The concept of a boundary layer is important in many problems in aerodynamics. The viscosity and fluid friction in the air is approximated as being significant only in this thin layer. This assumption makes the description of such aerodynamics much more tractable mathematically.

### Turbulence

In aerodynamics, turbulence is characterized by chaotic property changes in the flow. These include low momentum diffusion, high momentum convection, and rapid variation of pressure and flow velocity in space and time. Flow that is not turbulent is called laminar flow.

## Aerodynamic Force

The force exerted on a body whenever there is a relative velocity between the body and the air. There are only two basic sources of aerodynamic force: the pressure distribution and the frictional shear stress distribution exerted by the airflow on the body surface. The pressure exerted by the air at a point on the surface acts perpendicular to the surface at that point; and the shear stress, which is due to the frictional action of the air rubbing against the surface, acts tangentially to the surface at that point. The distribution of pressure and shear stress represent a distributed load over the surface. The net aerodynamic force on the body is due to the net imbalance between these distributed loads as they are summed (integrated) over the entire surface.

For purposes of discussion, it is convenient to consider the aerodynamic force on an airfoil. The net resultant aerodynamic force $R$ acting through the center of pressure on the airfoil represents mechanically the same effect as that due to the actual pressure and shear stress loads distributed over the body surface. The velocity of the airflow $V_\infty$ is called the free-stream velocity or the free-stream relative wind. By definition, the component of $R$ perpendicular to the relative wind is the lift, $L$, and the component of $R$ parallel to the relative wind is the drag $D$. The orientation of the body with respect to the direction of the free stream is given by the angle of attack, α. The magnitude of the aerodynamic force $R$ is governed by the density & rgr;$_\infty$ and velocity of the free stream, the size of the body, and the angle of attack.

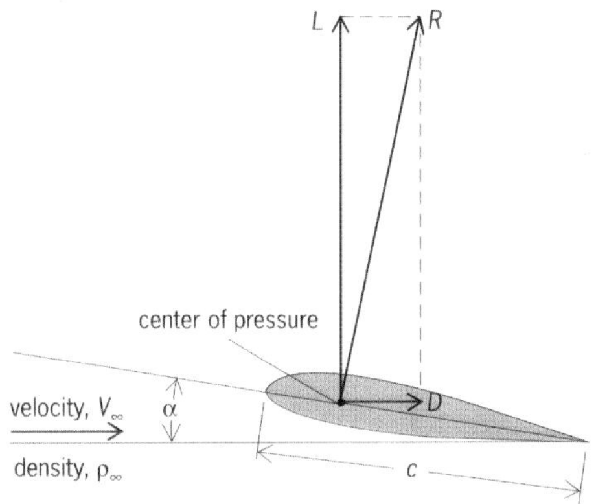


Resultant aerodynamic force ($R$), and its resolution into lift ($L$) and drag ($D$) components

An important measure of aerodynamic efficiency is the ratio of lift to drag, $L/D$. The higher the value of $L/D$, the more efficient is the lifting action of the body. The value of $L/D$ reaches a maximum, denoted by $(L/D)_{max}$, at a relatively low angle of attack. Beyond a certain angle the lift decreases with increasing α. In this region, the wing is said to be stalled. In the stall region the flow has separated from the top surface of the wing, creating a type of slowly recirculating dead-air region, which decreases the lift and substantially increases the drag.

## Lift Force

Lift is the force that directly opposes the weight of an airplane and holds the airplane in the air. Lift is generated by every part of the airplane, but most of the lift on a normal airliner is generated by the wings. Lift is a mechanical aerodynamic force produced by the motion of the airplane through the air. Because lift is a force, it is a vector quantity, having both a magnitude and a direction associated with it. Lift acts through the center of pressure of the object and is directed perpendicular to the flow direction. There are several factors which affect the magnitude of lift.

### How is Lift Generated?

There are many explanations for the generation of lift found in encyclopedias, in basic physics textbooks, and on Web sites. Unfortunately, many of the explanations are misleading and incorrect. Theories on the generation of lift have become a source of great controversy and a topic for heated arguments. To help you understand lift and its origins, a series of pages will describe the various theories and how some of the popular theories fail.

Lift occurs when a moving flow of gas is turned by a solid object. The flow is turned in one direction, and the lift is generated in the opposite direction, according to Newton's Third Law of action and reaction. Because air is a gas and the molecules are free to move about, any solid surface can deflect a flow. For an aircraft wing, both the upper and lower surfaces contribute to the flow turning. Neglecting the upper surface's part in turning the flow leads to an incorrect theory of lift.

## No Fluid, No Lift

Lift is a mechanical force. It is generated by the interaction and contact of a solid body with a fluid (liquid or gas). It is not generated by a force field, in the sense of a gravitational field,or an electromagnetic field, where one object can affect another object without being in physical contact. For lift to be generated, the solid body must be in contact with the fluid: no fluid, no lift. The Space Shuttle does not stay in space because of lift from its wings but because of orbital mechanics related to its speed. Space is nearly a vacuum. Without air, there is no lift generated by the wings.

## No Motion, No Lift

Lift is generated by the difference in velocity between the solid object and the fluid. There must be motion between the object and the fluid: no motion, no lift. It makes no difference whether the object moves through a static fluid, or the fluid moves past a static solid object. Lift acts perpendicular to the motion. Drag acts in the direction opposed to the motion.

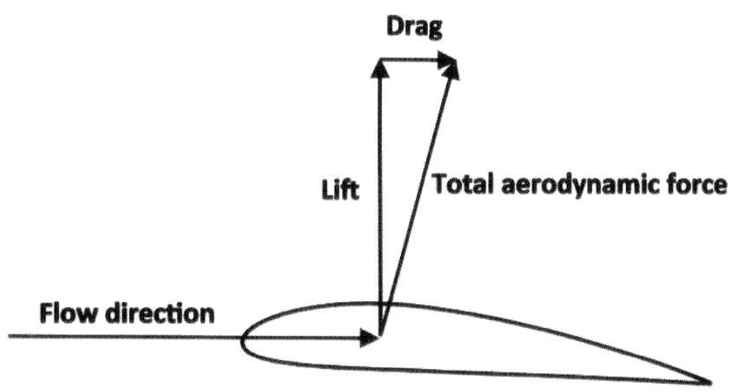


Lift is defined as the component of the total aerodynamic force perpendicular to the flow direction, and drag is the component parallel to the flow direction.

A fluid flowing past the surface of a body exerts a force on it. It makes no difference whether the fluid is flowing past a stationary body or the body is moving through a stationary volume of fluid. Lift is the component of this force that is perpendicular to the oncoming flow direction. Lift is always accompanied by a drag force, which is the component of the surface force parallel to the flow direction.

Lift is most commonly associated with the wings of fixed-wing aircraft, although it is more generally generated by many other streamlined bodies such as propellers, kites, helicopter rotors, racing car wings, maritime sails, and wind turbines in air, and by sailboat keels, ship's rudders, and hydrofoils in water. Lift is also exploited in the animal world, especially by birds, bats, and insects, and even in the plant world by the seeds of certain trees.

While the common meaning of the word "lift" assumes that lift opposes weight, lift can be in any direction with respect to gravity, since it is defined with respect to the direction of flow rather than to the direction of gravity. When an aircraft is cruising in straight and level flight, most of the lift opposes gravity. However, when an aircraft is climbing, descending, or banking in a turn the lift is tilted with respect to the vertical. Lift may also act as downforce in some aerobatic manoeuvres, or on the wing on a racing car. Lift may also be largely horizontal, for instance on a sailing ship.

The Lift discussed in this article is mainly in relation to airfoils, although marine hydrofoils and propellers share the same physical principles and work in the same way, despite differences between air and water such as density, compressibility, and viscosity.

## Simplified Physical Explanations of Lift on an Airfoil

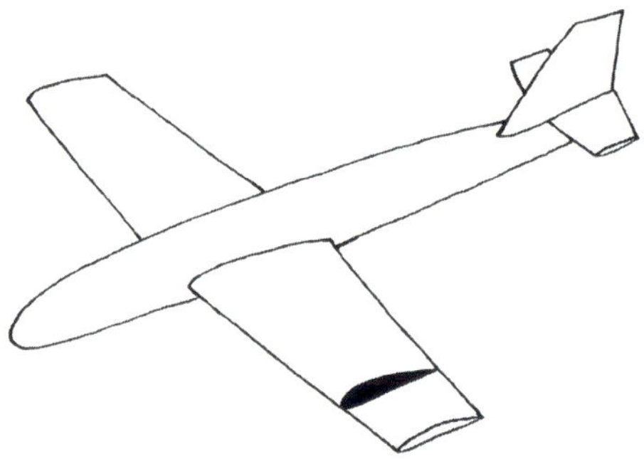

A cross-section of a wing defines an airfoil shape.

An airfoil is a streamlined shape that is capable of generating significantly more lift than drag. A flat plate can generate lift, but not as much as a streamlined airfoil, and with somewhat higher drag.

There are several ways to explain how an airfoil generates lift. Some are more complicated or more physically rigorous than others; some have been shown to be incorrect. For example, there are explanations based directly on Newton's laws of motion and explanations based on Bernoulli's principle. Either can be used to explain lift.

## Flow Deflection and Newton's Laws

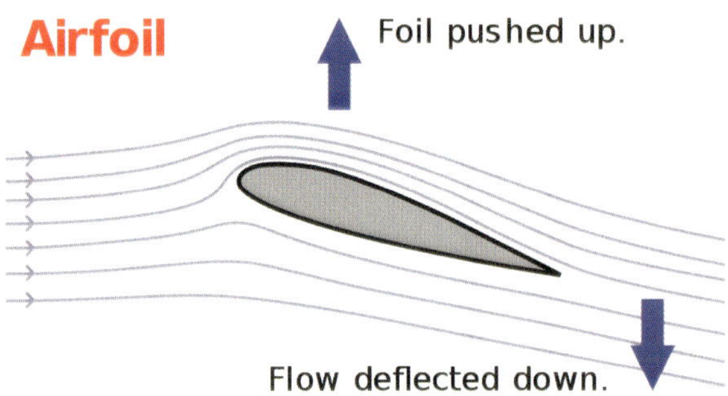


When an airfoil deflects air downwards, Newton's third law requires that the air must exert an equal upward reaction on the airfoil.

An airfoil generates lift by exerting a downward force on the air as it flows past. According to Newton's third law, the air must exert an equal and opposite (upward) force on the airfoil, which is Lift.

The airflow changes direction as it passes the airfoil and follows a path that is curved downward. According to Newton's second law, this change in flow direction requires a downward force applied to the air by the airfoil. Then Newton's third law requires the air to exert an upward force on the airfoil; thus a reaction force, Lift, is generated opposite to the directional change. In the case of an airplane wing, the wing exerts a downward force on the air and the air exerts an upward force on the wing.

The downward turning of the flow is not produced solely by the lower surface of the airfoil, and the air flow above the airfoil accounts for much of the downward-turning action.

## Increased Flow Speed and Bernoulli's Principle

Bernoulli's principle states: "In a flow where no energy is being added or taken away, the sum of its various energies is a constant: consequently where the velocity increases the pressure decreases and vice versa." Thus, there is a direct mathematical relationship between the pressure and the speed, so if one knows the speed at all points within the airflow one can calculate the pressure, and vice versa. For any airfoil generating lift, there must be a pressure imbalance, i.e. lower average air pressure on the top than on the bottom. Bernoulli's principle states that this pressure difference must be accompanied by a speed difference.

## Conservation of Mass

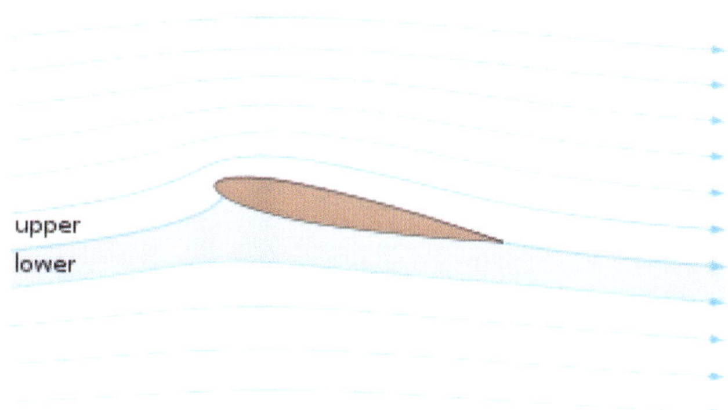


Streamlines and streamtubes around an airfoil generating lift. Note the narrower streamtubes above and the wider streamtubes below.

Starting with the flow pattern observed in both theory and experiments, the increased flow speed over the upper surface can be explained in terms of streamtube pinching and conservation of mass.

For incompressible flow, the rate of volume flow (e.g. volume units per minute) must be constant within each streamtube since matter is not created or destroyed. If a streamtube becomes narrower, the flow speed must increase in the narrower region to maintain the constant flow rate, to satisfy the principle of conservation of mass.

The upper streamtubes constrict as they flow up and around the airfoil. Conservation of mass says that the flow speed must increase as the stream tube area decreases. Similarly, the lower streamtubes expand and their flowrate slows.

From Bernoulli's principle, the pressure on the upper surface where the flow is moving faster is lower than the pressure on the lower surface where it is moving slower. This pressure difference creates a net aerodynamic force, pointing upward.

## Limitations of the Simplified Explanations

Producing a lift force requires maintaining pressure differences in both the vertical and horizontal directions, and thus requires both downward turning of the flow and changes in flow speed consistent with Bernoulli's principle. The simplified explanations given above are therefore incomplete because they define lift in terms of only one or the other. And depending on the details, they have other shortcomings as well.

The explanation based on Flow deflection and Newton's laws is correct but is incomplete. It does not explain how the airfoil can impart downward turning to a much deeper swath of the flow than it actually touches. Further, it doesn't explain how the pressure differences in the horizontal direction are sustained. That is, it leaves out the Bernoulli part of the interaction.

Explanations based on Increased flow speed and Bernoulli's principle first try to establish that there is higher flow speed over the upper surface, but they fail to explain correctly what causes the flow to speed up:

- The Conservation of mass explanation that relies on narrowing of the streamtubes over the upper surface does not explain why the streamtubes change size. To see why the air flows the way it does requires more sophisticated analysis.
- Sometimes a geometrical argument is offered to demonstrate why the streamtubes change size: it is asserted that the top "obstructs" or "constricts" the air more than the bottom, hence narrower streamtubes. For conventional wings that are flat on the bottom and curved on top this makes some intuitive sense. But it does not explain how flat plates, symmetric airfoils, sailboat sails, or conventional airfoils flying upside down can generate lift, and attempts to calculate lift based on the amount of constriction do not predict experimental results.
- A common version that relies on equal-transit-time is simply wrong, as explained below under False explanation based on equal transit-time.

Bernoulli-only explanations imply that a speed difference arises from causes other than a pressure difference, and that the speed difference then leads to a pressure difference by Bernoulli's principle. This implied one-way causation is a misconception. The real cause-and-effect relationship between pressure and velocity is reciprocal. Finally, Bernoulli-only explanations don't explain how the pressure differences in the vertical direction are sustained. That is, they leave out the downward-turning part of the interaction.

## Basic Attributes of Lift

Lift is a result of pressure differences and depends on angle of attack, airfoil shape, air density, and airspeed.

## Pressure Differences

Pressure is the normal force per unit area exerted by the air on itself and on surfaces that it touches. The lift force is transmitted through the pressure, which acts perpendicular to the surface of the airfoil. The air maintains physical contact at all points. Thus, the net force manifests itself as pressure differences. The direction of the net force implies that the average pressure on the upper surface of the airfoil is lower than the average pressure on the underside.

These pressure differences arise in conjunction with the curved airflow. When a fluid follows a curved path, there is a pressure gradient perpendicular to the flow direction with higher pressure on the outside of the curve and lower pressure on the inside. This direct relationship between curved streamlines and pressure differences, sometimes called the streamline curvature theorem, was derived from Newton's second law by Leonhard Euler in 1754:

$$\frac{\mathrm{d}p}{\mathrm{d}R} = \rho\frac{v^2}{R}$$

The left side of this equation represents the pressure difference perpendicular to the fluid flow. On the right hand side ρ is the density, v is the velocity, and R is the radius of curvature. This formula shows that higher velocities and tighter curvatures create larger pressure differentials and that for straight flow ($R \to \infty$) the pressure difference is zero.

## Angle of Attack

Angle of attack of an airfoil

The angle of attack is the angle between the chord line of an airfoil and the oncoming airflow. A symmetrical airfoil will generate zero lift at zero angle of attack. But as the angle of attack increases, the air is deflected through a larger angle and the vertical component of the airstream velocity increases, resulting in more lift. For small angles a symmetrical airfoil will generate a lift force roughly proportional to the angle of attack.

As the angle of attack increases, the lift reaches a maximum at some angle; increasing the angle of attack beyond this critical angle of attack causes the upper-surface flow to separate from the wing; there is less deflection downward so the airfoil generates less lift. The airfoil is said to be stalled.

## Airfoil Shape

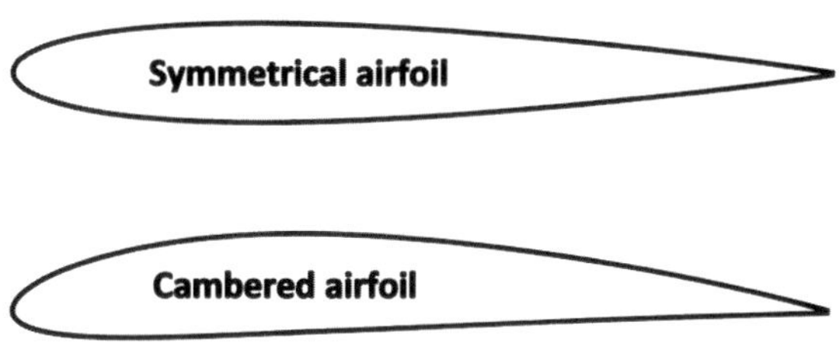


An airfoil with camber compared to a symmetrical airfoil

The lift force depends on the shape of the airfoil, especially the amount of camber (curvature such that the upper surface is more convex than the lower surface, as illustrated at right). Increasing the camber generally increases lift.

Cambered airfoils will generate lift at zero angle of attack. When the chord line is horizontal, the trailing edge has a downward direction and since the air follows the trailing edge it is deflected downward. When a cambered airfoil is upside down, the angle of attack can be adjusted so that the lift force is upwards. This explains how a plane can fly upside down.

## Flow Conditions

The ambient flow conditions which affect lift include the fluid density, viscosity and speed of flow. Density is affected by temperature, and by the medium's acoustic velocity - i.e. by compressibility effects.

## Air Speed and Density

Lift is proportional to the density of the air and approximately proportional to the square of the flow speed. Lift also depends on the size of the wing, being generally proportional to the wing's area projected in the lift direction. In calculations it is convenient to quantify lift in terms of a "Lift coefficient" based on these factors.

## Boundary Layer and Profile Drag

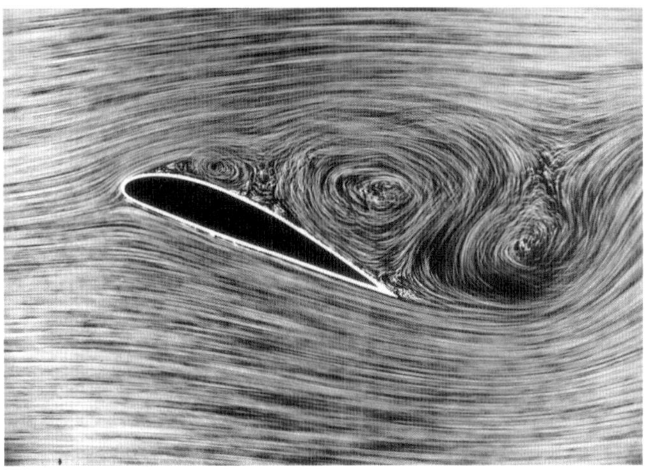

Airflow separating from a wing at a high angle of attack

No matter how smooth the surface of an airfoil seems, any surface is rough on the scale of air molecules. Air molecules flying into the surface bounce off the rough surface in random directions superposed on their original velocity. The result is that when the air is viewed as if it were a continuous material, it is seen to be unable to slide along the surface, and the air's velocity relative to the airfoil decreased to nearly zero at the surface. goes to practically zero, something known as the no-slip condition. Because the air at the surface has near-zero velocity, and air away from the surface is moving, there is a thin boundary layer in which air close to the surface is subjected to a shearing motion. The air's viscosity resists the shearing, giving rise to a shear stress at the airfoil's surface called skin-friction drag. Over most of the surface of most airfoils, the boundary layer is naturally turbulent, which increases skin-friction drag.

Under usual flight conditions, the boundary layer remains attached to both the upper and lower surfaces all the way to the trailing edge, and its effect on the rest of the flow is modest. Compared to the predictions of inviscid-flow theory, in which there is no boundary layer, the attached boundary layer reduces the lift by a modest amount and modifies the pressure distribution somewhat, which results in a viscosity-related pressure drag over and above the skin-friction drag. The total of the skin-friction drag and the viscosity-related pressure drag is usually called the profile drag.

## Stalling

An airfoil's maximum lift at a given airspeed is limited by boundary-layer separation. As the angle of attack is increased, a point is reached where the boundary layer can no longer remain attached to the upper surface. When the boundary layer separates, it leaves a region of recirculating flow above the upper surface, as illustrated in the flow-visualization photo at right. This is known as the *stall*, or *stalling*. At angles of attack above the stall, lift is significantly reduced, though it does not drop to zero. The maximum lift that can be achieved before stall, in terms of the lift coefficient, is generally less than 1.5 for single-element airfoils and can be more than 3.0 for airfoils with high-lift slotted flaps and leading-edge devices deployed.

## Bluff Bodies

The flow around bluff bodies – i.e. without a streamlined shape, or stalling airfoils – may also generate lift, in addition to a strong drag force. This lift may be steady, or it may oscillate due to vortex shedding. Interaction of the object's flexibility with the vortex shedding may enhance the effects of fluctuating lift and cause vortex-induced vibrations. For instance, the flow around a circular cylinder generates a Kármán vortex street: vortices being shed in an alternating fashion from the cylinder's sides. The oscillatory nature of the flow produces a fluctuating lift force on the cylinder, even though the net (mean) force is negligible. The lift force frequency is characterised by the dimensionless Strouhal number, which depends on the Reynolds number of the flow.

For a flexible structure, this oscillatory lift force may induce vortex-induced vibrations. Under certain conditions – for instance resonance or strong spanwise correlation of the lift force – the resulting motion of the structure due to the lift fluctuations may be strongly enhanced. Such vibrations may pose problems and threaten collapse in tall man-made structures like industrial chimneys.

In the Magnus effect, a lift force is generated by a spinning cylinder in a freestream. Here the mechanical rotation acts on the boundary layer, causing it to separate at different locations on the two sides of the cylinder. The asymmetric separation changes the effective shape of the cylinder as far as the flow is concerned such that the cylinder acts like a lifting airfoil with circulation in the outer flow.

## A More Comprehensive Physical Explanation

There are two main popular explanations: one based on downward deflection of the flow (Newton's laws), and one based on pressure differences accompanied by changes in flow speed (Bernoulli's principle). Either of these, by itself, correctly identifies some aspects of the lifting flow but leaves other important aspects of the phenomenon unexplained. A more comprehensive explanation involves both downward deflection and pressure differences (including changes in flow speed associated with the pressure differences).

## Lift at the Airfoil Surface

The airfoil shape and angle of attack work together so that the airfoil exerts a downward force on the air as it flows past. According to Newton's third law, the air must then exert an equal and opposite (upward) force on the airfoil, which is the lift.

The net force exerted by the air occurs as a pressure difference over the airfoil's surfaces. Pressure in a fluid is always positive in an absolute sense, so that pressure must always be thought of as pushing, and never as pulling. The pressure thus pushes inward on the airfoil everywhere on both the upper and lower surfaces. The flowing air reacts to the presence of the wing by reducing the pressure on the wing's upper surface and increasing the pressure on the lower surface. The pressure on the lower surface pushes up harder than the reduced pressure on the upper surface pushes down, and the net result is upward lift.

The pressure difference which results in Lift acts directly on the airfoil surfaces; however, understanding how the pressure difference is produced requires understanding what the flow does over a wider area.

## The Wider Flow around the Airfoil

An airfoil affects the speed and direction of the flow over a wide area, producing a pattern called a *velocity field*. When an airfoil produces lift, the flow ahead of the airfoil is deflected upward, the flow above and below the airfoil is deflected downward, and the flow behind the airfoil is deflected upward again, leaving the air far behind the airfoil in the same state as the oncoming flow far ahead. The flow above the upper surface is sped up, while the flow below the airfoil is slowed down. Together with the upward deflection of air in front and the downward deflection of the air immediately behind, this establishes a net circulatory component of the flow. The downward deflection and the changes in flow speed are pronounced and extend over a wide area, as can be seen in the image below. These differences in the direction and speed of the flow are greatest close to the airfoil and decrease gradually far above and below. All of these features of the velocity field also appear in theoretical models for lifting flows.

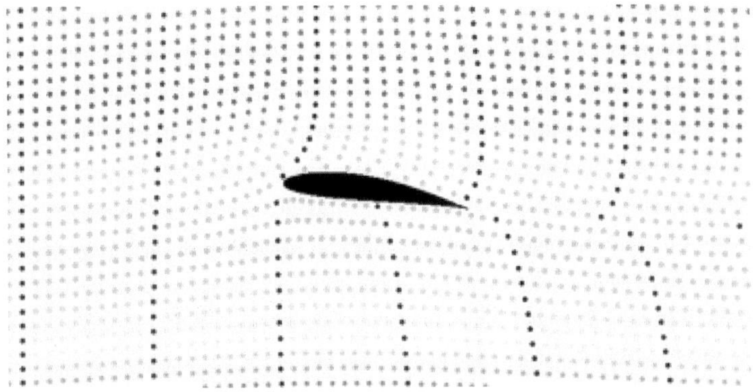

Flow around an airfoil: the dots move with the flow

The black dots are on time slices, which split into two – an upper and lower part – at the leading edge. A marked speed difference between the upper-and lower-surface streamlines is shown most clearly in the image, with the upper markers arriving at the trailing edge long before the lower ones. Colors of the dots indicate streamlines.

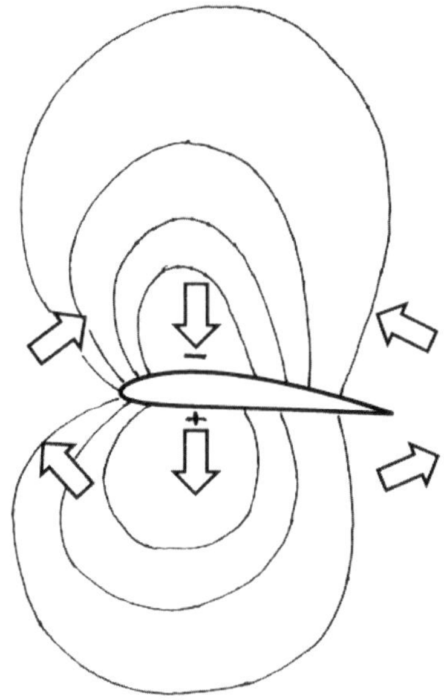


Pressure distribution with isobars around a lifting airfoil. The plus sign indicates pressure higher than ambient, and the minus sign indicates pressure lower than ambient (NOT negative pressure in the absolute sense). The block arrows indicate the directions of net forces on fluid parcels in different parts of the flowfield.

The pressure is also affected over a wide area, in a pattern of non-uniform pressure called a *pressure field*. When an airfoil produces lift, there is a diffuse region of low pressure above the airfoil, and usually a diffuse region of high pressure below, as illustrated by the isobars (curves of constant pressure) in the drawing. The pressure difference that acts on the surface is just part of this pressure field.

## Mutual Interaction of Pressure Differences and Changes in Flow Velocity

The non-uniform pressure exerts forces on the air in the direction from higher pressure to lower pressure. The direction of the force is different at different locations around the airfoil, as indicated by the block arrows in the *pressure distribution with isobars* figure. Air above the airfoil is pushed toward the center of the low-pressure region, and air below the airfoil is pushed outward from the center of the high-pressure region.

According to *Newton's second law*, a force causes air to accelerate in the direction of the force. Thus the vertical arrows in the *pressure distribution with isobars* figure indicate that air above and below the airfoil is accelerated, or turned downward, and that the non-uniform pressure is thus the cause of the downward deflection of the flow visible in the flow animation. To produce this downward turning, the airfoil must have a positive angle of attack or have its rear portion curved downward as on an airfoil with camber. Note that the downward turning of the flow over the upper surface is the result of the air being pushed downward by higher pressure above it than below it. Some explanations that refer to the "Coandă effect" suggest that viscosity plays a key role in the downward turning, but this is false.

The arrows ahead of the airfoil indicate that the flow ahead of the airfoil is deflected upward, and the arrows behind the airfoil indicate that the flow behind is deflected upward again, after being deflected downward over the airfoil.

The arrows ahead of the airfoil and behind also indicate that air passing through the low-pressure region above the airfoil is sped up as it enters, and slowed back down as it leaves. Air passing through the high-pressure region below the airfoil sees the opposite - it is slowed down and then sped up. Thus the non-uniform pressure is also the cause of the changes in flow speed visible in the flow animation. The changes in flow speed are consistent with *Bernoulli's principle*, which states that in a steady flow without viscosity, lower pressure means higher speed, and higher pressure means lower speed.

Thus changes in flow direction and speed are directly caused by the non-uniform pressure. But this cause-and-effect relationship is not just one-way; it works in both directions simultaneously. The air's motion is affected by the pressure differences, but the existence of the pressure differences depends on the air's motion. The relationship is thus a mutual, or reciprocal, interaction: Air flow changes speed or direction in response to pressure differences, and the pressure differences are sustained by the air's resistance to changing speed or direction. A pressure difference can exist only if something is there for it to push against. In aerodynamic flow, the pressure difference pushes against the air's inertia, as the air is accelerated by the pressure difference. This is why the air's mass is part of the calculaqtion, and why Lift depends on air density.

Sustaining the pressure difference that exerts the lift force on the airfoil surfaces requires sustaining a pattern of non-uniform pressure in a wide area around the airfoil. This requires maintaining pressure differences in both the vertical and horizontal directions, and thus requires both downward turning of the flow and changes in flow speed according to Bernoulli's principle. The pressure differences and the changes in flow direction and speed sustain each other in a mutual interaction. The pressure differences follow naturally from Newton's second law and from the fact that flow along the surface follows the predominantly downward-sloping contours of the airfoil. And the fact that the air has mass is crucial to the interaction.

## Simpler Explanations

Producing a lift force requires both downward turning of the flow and changes in flow speed consistent with Bernoulli's principle. Each of the simplified explanations given above in Simplified physical explanations of lift on an airfoil falls short by trying to explain lift in terms of only one or the other, thus explaining only part of the phenomenon and leaving other parts unexplained.

## Quantifying Lift

### Pressure Integration

When the pressure distribution on the airfoil surface is known, determining the total lift requires adding up the contributions to the pressure force from local elements of the surface, each with its own local value of pressure. The total lift is thus the integral of the pressure, in the direction perpendicular to the farfield flow, over the airfoil surface.

$$L = \oint p\mathrm{n} \cdot \mathrm{k}\, dS,$$

where:

- S is the projected (planform) area of the airfoil, measured normal to the mean airflow;

- n is the normal unit vector pointing into the wing;
- k is the vertical unit vector, normal to the freestream direction.

The above lift equation neglects the skin friction forces, which are small compared to the pressure forces.

By using the streamwise vector i parallel to the freestream in place of k in the integral, we obtain an expression for the pressure drag $D_p$ (which includes the pressure portion of the profile drag and, if the wing is three-dimensional, the induced drag). If we use the spanwise vector j, we obtain the side force $Y$.

$$D_p = \oint p\mathrm{n} \cdot \mathrm{i}\, dS,$$

$$Y = \oint p\mathrm{n} \cdot \mathrm{j}\, dS.$$

The validity of this integration generally requires the airfoil shape to be a closed curve that is piece-wise smooth.

## Lift Coefficient

Lift depends on the size of the wing, being approximately proportional to the wing area. It is often convenient to quantify the lift of a given airfoil by its *lift coefficient* $C_L$, which defines its overall lift in terms of a unit area of the wing.

If the value of $C_L$ for a wing at a specified angle of attack is given, then the lift produced for specific flow conditions can be determined:

$$L = \tfrac{1}{2}\rho v^2 S C_L$$

where

- $L$ is the lift force
- $\rho$ is the air density
- $v$ is the velocity or true airspeed
- $S$ is the planform (projected) wing area
- $C_L$ is the lift coefficient at the desired angle of attack, Mach number, and Reynolds number

## Three-dimensional Flow

The flow around a three-dimensional wing involves significant additional issues, especially relating to the wing tips. For a wing of low aspect ratio, such as a typical delta wing, two-dimensional thories may provide a poor model and three-dimensional flow effects can dominate. Even for wings of high aspect ratio, the three-dimensional effects associated with finite span can affect the whole span, not just close to the tips.

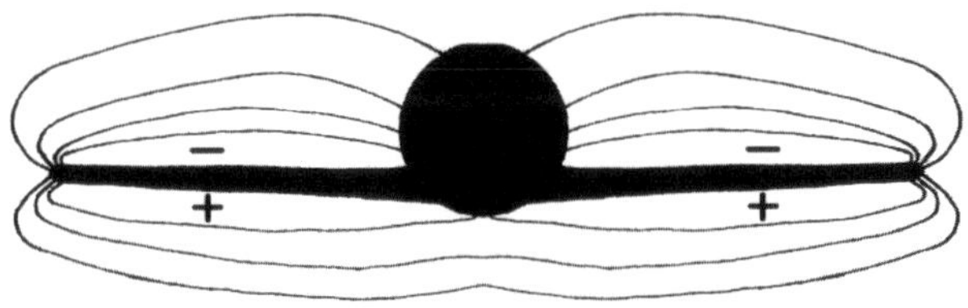

Cross-section of an airplane wing-body combination showing the isobars of the three-dimensional lifting flow

Cross-section of an airplane wing-body combination showing velocity vectors of the three-dimensional lifting flow

## Wing Tips and Spanwise Distribution

The vertical pressure gradient at the wing tips causes air to flow sideways, out from under the wing then up and back over the upper surface. This reduces the pressure gradient at the wing tip, therefore also reducing lift. The lift tends to decrease in the spanwise direction from root to tip, and the pressure distributions around the airfoil sections change accordingly in the spanwise direction. Pressure distributions in planes perpendicular to the flight direction tend to look like the illustration at right. This spanwise-varying pressure distribution is sustained by a mutual interaction with the velocity field. Flow below the wing is accelerated outboard, flow outboard of the tips is accelerated upward, and flow above the wing is accelerated inboard, which results in a flow pattern.

There is more downward turning of the flow than there would be in a two-dimensional flow with the same airfoil shape and sectional lift, and a higher sectional angle of attack is required to achieve the same lift compared to a two-dimensional flow. The wing is effectively flying in a downdraft of its own making, as if the freestream flow were tilted downward, with the result that the total aerodynamic force vector is tilted backward slightly compared to what it would be in two dimensions. The additional backward component of the force vector is called lift-induced drag.

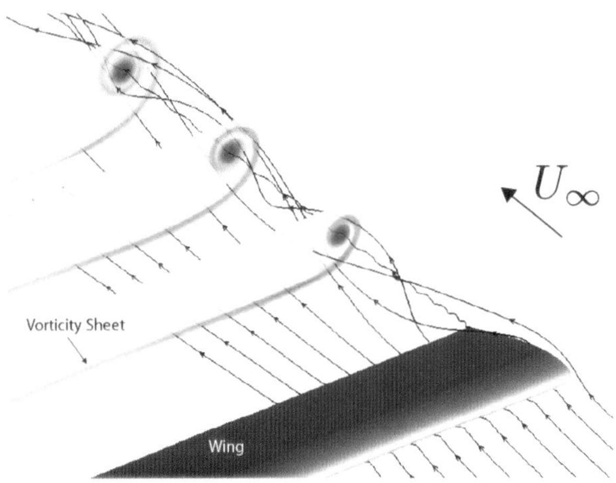

Euler computation of a tip vortex rolling up from the trailed vorticity sheet

The difference in the spanwise component of velocity above and below the wing (between being in the inboard direction above and in the outboard direction below) persists at the trailing edge and into the wake downstream. After the flow leaves the trailing edge, this difference in velocity takes place across a relatively thin shear layer called a vortex sheet.

## Horseshoe Vortex System

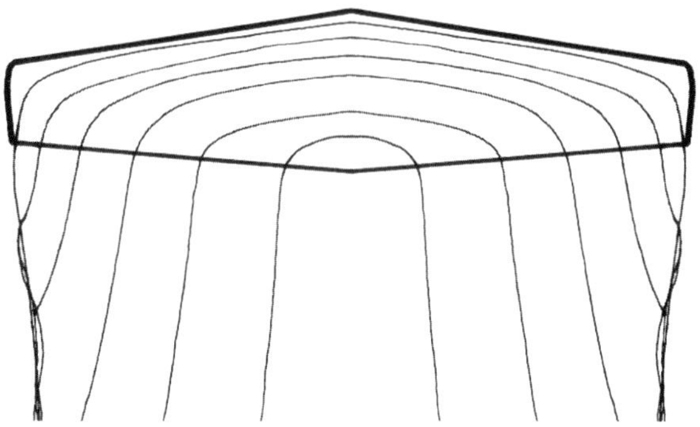

Planview of a wing showing the horseshoe vortex system

The wingtip flow leaving the wing creates a tip vortex. As the main vortex sheet passes downstream from the trailing edge, it rolls up at its outer edges, merging with the tip vortices. The combination of the wingtip vortices and the vortex sheets feeding them is called the vortex wake.

In addition to the vorticity in the trailing vortex wake there is vorticity in the wing's boundary layer, called 'bound vorticity', which connects the trailing sheets from the two sides of the wing into a vortex system in the general form of a horseshoe. The horseshoe form of the vortex system was recognized by the British aeronautical pioneer Lanchester in 1907.

Given the distribution of bound vorticity and the vorticity in the wake, the Biot-Savart law (a vector-calculus relation) can be used to calculate the velocity perturbation anywhere in the field, caused by the lift on the wing. Approximate theories for the lift distribution and lift-induced drag of three-dimensional wings are based on such analysis applied to the wing's horseshoe vortex system. In these theories, the bound vorticity is usually idealized and assumed to reside at the camber surface inside the wing.

Because the velocity is deduced from the vorticity in such theories, some authors describe the situation to imply that the vorticity is the cause of the velocity perturbations, using terms such as "the velocity induced by the vortex," for example. But attributing mechanical cause-and-effect between the vorticity and the velocity in this way is not consistent with the physics. The velocity perturbations in the flow around a wing are in fact produced by the pressure field.

## Manifestations of Lift in the Farfield

### Integrated Force/Momentum Balance in Lifting Flows

The flow around a lifting airfoil must satisfy Newton's second law regarding conservation of momentum, both locally at every point in the flow field, and in an integrated sense over any extended region of the flow. For an extended region, Newton's second law takes the form of the *momentum theorem for a control volume*, where a control volume can be any region of the flow chosen for analysis. The momentum theorem states that the integrated force exerted at the boundaries of the control volume (a surface integral), is equal to the integrated time rate of change (material derivative) of the momentum of fluid parcels passing through the interior of the control volume. For a steady flow, this can be expressed in the form of the net surface integral of the flux of momentum through the boundary.

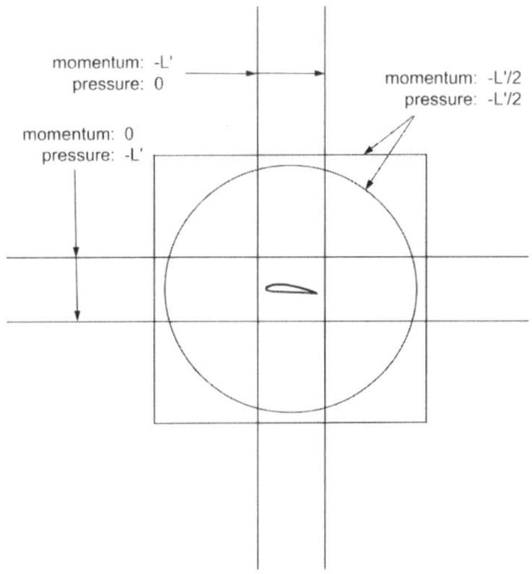


Control volumes of different shapes that have been used in analyzing the momentum balance in the 2D flow around a lifting airfoil. The airfoil is assumed to exert a downward force -L' per unit span on the air, and the proportions in which that force is manifested as momentum fluxes and pressure differences at the outer boundary are indicated for each different shape of control volume.

The lifting flow around a 2D airfoil is usually analyzed in a control volume that completely surrounds the airfoil, so that the inner boundary of the control volume is the airfoil surface, where the downward force per unit span $-L'$ is exerted on the fluid by the airfoil. The outer boundary is usually either a large circle or a large rectangle. At this outer boundary distant from the airfoil, the velocity and pressure are well represented by the velocity and pressure associated with a uniform flow plus a vortex, and viscous stress is negligible, so that the only force that must be integrated over the outer boundary is the pressure. The free-stream velocity is usually assumed to be horizontal, with lift vertically upward, so that the vertical momentum is the component of interest.

For the free-air case (no ground plane), the force $-L'$ exerted by the airfoil on the fluid is manifested partly as momentum fluxes and partly as pressure differences at the outer boundary, in proportions that depend on the shape of the outer boundary, as shown in the diagram at right. For a flat horizontal rectangle that is much longer than it is tall, the fluxes of vertical momentum through the front and back are negligible, and the lift is accounted for entirely by the integrated pressure differences on the top and bottom. For a square or circle, the momentum fluxes and pressure differences account for half the lift each. For a vertical rectangle that is much taller than it is wide, the unbalanced pressure forces on the top and bottom are negligible, and lift is accounted for entirely by momentum fluxes, with a flux of upward momentum that enters the control volume through the front accounting for half the lift, and a flux of downward momentum that exits the control volume through the back accounting for the other half.

### Lift Reacted by Overpressure on the Ground under an Airplane

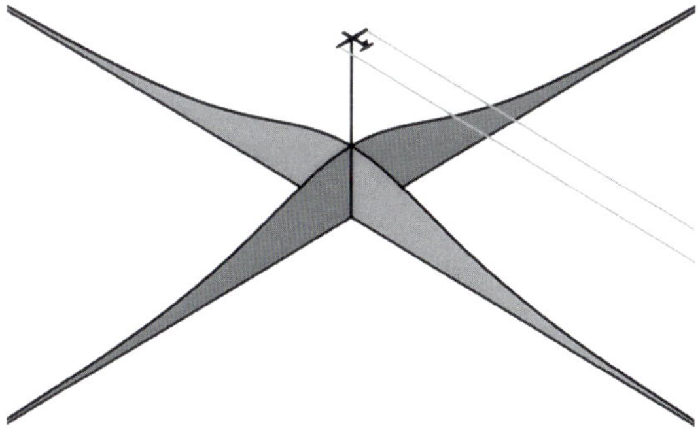

Illustration of the distribution of higher-than-ambient pressure on the ground under an airplane in flight

An airfoil produces a pressure field in the surrounding air, as explained under "The wider flow around the airfoil" above. The pressure differences associated with this field die off gradually, becoming very small at large distances, but never disappearing altogether. Below the airplane, the pressure field persists as a positive pressure disturbance that reaches the ground, forming a pattern of slightly-higher-than-ambient pressure on the ground, as shown on the right. Although the pressure differences are very small far below the airplane, they are spread over a wide area and add up to a substantial force. For steady, level flight, the integrated force due to the pressure differences is equal to the total aerodynamic lift of the airplane and to the airplane's weight. According to Newton's third law, this pressure force exerted on the ground by the air is matched by an equal-and-opposite upward force exerted on the air by the ground, which offsets all of the downward force exerted on the air by the airplane. The net force due to the lift, acting on the atmosphere as a whole, is therefore zero, and thus there is no integrated accumulation of vertical momentum in the atmosphere, as was noted by Lanchester early in the development of modern aerodynamics.

## Aerodynamic Drag

Drag is the aerodynamic force that opposes an aircraft's motion through the air. Drag is generated by every part of the airplane (even the engines). How is drag generated?

Drag is a mechanical force. It is generated by the interaction and contact of a solid body with a fluid (liquid or gas). It is not generated by a force field, in the sense of a gravitational field or an electromagnetic field, where one object can affect another object without being in physical contact. For drag to be generated, the solid body must be in contact with the fluid. If there is no fluid, there is no drag. Drag is generated by the difference in velocity between the solid object and the fluid. There must be motion between the object and the fluid. If there is no motion, there is no drag. It makes no difference whether the object moves through a static fluid or whether the fluid moves past a static solid object.

Drag is a force and is therefore a vector quantity having both a magnitude and a direction. Drag acts in a direction that is opposite to the motion of the aircraft. Lift acts perpendicular to the motion. There are many factors that affect the magnitude of the drag. Many of the factors also affect lift but there are some factors that are unique to aircraft drag.

The pressure distribution acting on a body's surface exerts normal forces on the body. Those forces can be summed and the component of that force that acts downstream represents the drag force, $D_{pr}$, due to pressure distribution acting on the body. The nature of these normal forces combines shock wave effects, vortex system generation effects, and wake viscous mechanisms.

The viscosity of the fluid has a major effect on drag. In the absence of viscosity, the pressure forces acting to retard the vehicle are canceled by a pressure force further aft that acts to push the vehicle forward; this is called pressure recovery and the result is that the drag is zero. That is to say, the work the body does on the airflow, is reversible and is recovered as there are no frictional effects to convert the flow energy into heat. Pressure recovery acts even in the case of viscous flow. Viscosity, however results in pressure drag and it is the dominant component of drag in the case of vehicles with regions of separated flow, in which the pressure recovery is fairly ineffective.

The friction drag force, which is a tangential force on the aircraft surface, depends substantially on boundary layer configuration and viscosity. The net friction drag, $D_f$, is calculated as the downstream projection of the viscous forces evaluated over the body's surface.

The sum of friction drag and pressure (form) drag is called viscous drag. This drag component is due to viscosity. In a thermodynamic perspective, viscous effects represent irreversible phenomena and, therefore, they create entropy. The calculated viscous drag $D_v$ use entropy changes to accurately predict the drag force.

When the airplane produces lift, another drag component results. Induced drag, symbolized $D_i$, is due to a modification of the pressure distribution due to the trailing vortex system that accompanies the lift production. An alternative perspective on lift and drag is gained from considering the change of momentum of the airflow. The wing intercepts the airflow and forces the flow to move downward. This results in an equal and opposite force acting upward on the wing which is the lift force. The change of momentum of the airflow downward results in a reduction of the rearward momentum of the flow which is the result of a force acting forward on the airflow and applied by the wing to the air flow; an equal but opposite force acts on the wing rearward which is the induced drag. Induced drag tends to be the most important component for airplanes during take-off or landing flight. Another drag component, namely wave drag, $D_w$, results from shock waves in transonic and supersonic flight speeds. The shock waves induce changes in the boundary layer and pressure distribution over the body surface.

## Lift-induced Drag

Lift-induced drag (also called induced drag) is drag which occurs as the result of the creation of lift on a three-dimensional lifting body, such as the wing or fuselage of an airplane. Induced drag consists primarily of two components: drag due to the creation of trailing vortices (vortex drag); and the presence of additional viscous drag (lift-induced viscous drag) that is not present when lift is zero. The trailing vortices in the flow-field, present in the wake of a lifting body, derive from the turbulent mixing of air from above and below the body which flows in slightly different directions as a consequence of creation of lift.

With other parameters remaining the same, as the lift generated by a body increases, so does the lift-induced drag. This means that as the wing's angle of attack increases (up to a maximum called the stalling angle), the lift coefficient also increases, and so too does the lift-induced drag. At the onset of stall, lift is abruptly decreased, as is lift-induced drag, but viscous pressure drag, a component of parasite drag, increases due to the formation of turbulent unattached flow in the wake behind the body.

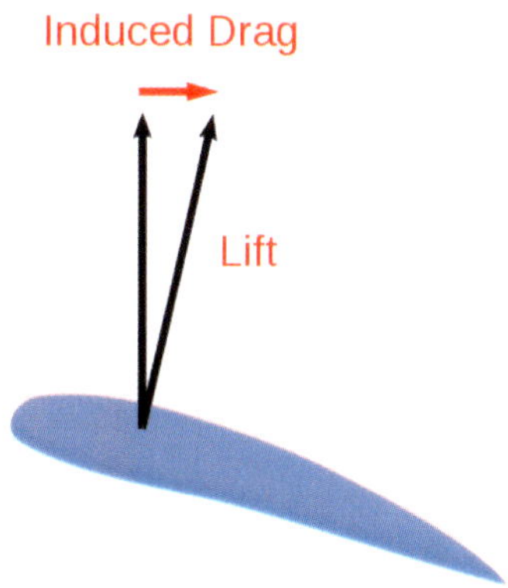


Induced drag *vs.* lift

## Parasitic Drag

Parasitic drag is drag caused by moving a solid object through a fluid. Parasitic drag is made up of multiple components including viscous pressure drag (form drag), and drag due to surface roughness (skin friction drag). Additionally, the presence of multiple bodies in relative proximity may incur so called interference drag, which is sometimes described as a component of parasitic drag.

In aviation, induced drag tends to be greater at lower speeds because a high angle of attack is required to maintain lift, creating more drag. However, as speed increases the angle of attack can be reduced and the induced drag decreases. Parasitic drag, however, increases because the fluid is flowing more quickly around protruding objects increasing friction or drag. At even higher speeds (transonic), wave drag enters the picture. Each of these forms of drag changes in proportion to the others based on speed. The combined overall drag curve therefore shows a minimum at some airspeed - an aircraft flying at this speed will be at or close to its optimal efficiency. Pilots will use this speed to maximize endurance (minimum fuel consumption), or maximize gliding range in the event of an engine failure.

## Power Curve in Aviation

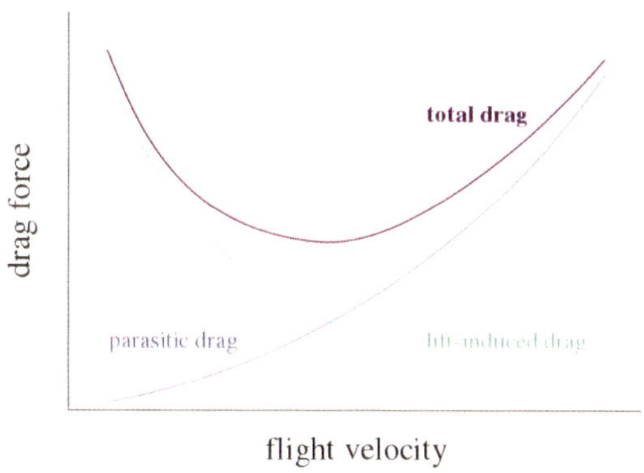


The *power curve*: form and induced drag *vs.* airspeed

The interaction of parasitic and induced drag *vs.* airspeed can be plotted as a characteristic curve, illustrated here. In aviation, this is often referred to as the *power curve*, and is important to pilots because it shows that, below a certain airspeed, maintaining airspeed counterintuitively requires *more* thrust as speed decreases, rather than less. The consequences of being "behind the curve" in flight are important and are taught as part of pilot training. At the subsonic airspeeds where the "U" shape of this curve is significant, wave drag has not yet become a factor, and so it is not shown in the curve.

## Wave Drag in Transonic and Supersonic Flow

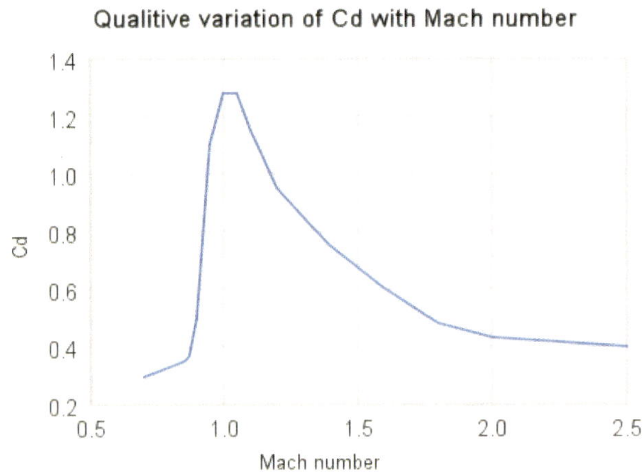


Qualitative variation in Cd factor with Mach number for aircraft

Wave drag (also called compressibility drag) is drag that is created when a body moves in a compressible fluid and at speeds that are close to the speed of sound in that fluid. In aerodynamics, wave drag consists of multiple components depending on the speed regime of the flight.

In transonic flight (Mach numbers greater than about 0.8 and less than about 1.4), wave drag is the result of the formation of shockwaves in the fluid, formed when local areas of supersonic (Mach number greater than 1.0) flow are created. In practice, supersonic flow occurs on bodies traveling well below the speed of sound, as the local speed of air increases as it accelerates over the body to speeds above Mach 1.0. However, full supersonic flow over the vehicle will not develop until well past Mach 1.0. Aircraft flying at transonic speed often incur wave drag through the normal course of operation. In transonic flight, wave drag is commonly referred to as transonic compressibility drag. Transonic compressibility drag increases significantly as the speed of flight increases towards Mach 1.0, dominating other forms of drag at those speeds.

In supersonic flight (Mach numbers greater than 1.0), wave drag is the result of shockwaves present in the fluid and attached to the body, typically oblique shockwaves formed at the leading and trailing edges of the body. In highly supersonic flows, or in bodies with turning angles sufficiently large, unattached shockwaves, or bow waves will instead form. Additionally, local areas of transonic flow behind the initial shockwave may occur at lower supersonic speeds, and can lead to the development of additional, smaller shockwaves present on the surfaces of other lifting bodies, similar to those found in transonic flows. In supersonic flow regimes, wave drag is commonly separated into two components, supersonic lift-dependent wave drag and supersonic volume-dependent wave drag.

The closed form solution for the minimum wave drag of a body of revolution with a fixed length was found by Sears and Haack, and is known as the Sears-Haack Distribution. Similarly, for a fixed volume, the shape for minimum wave drag is the Von Karman Ogive.

Busemann's Biplane is not, in principle, subject to wave drag at all when operated at its design speed, but is incapable of generating lift.

## Aircraft Drag Reduction

Every aircraft builder is interested in improving the speed of his airplane, or at least reducing its fuel consumption so that he/she gets the highest mileage out of every gallon or liter of fuel in the tanks or the lowest hourly rate.

Basically the form of the aircraft defines how much drag it generates and, in unaccelerated flight, how much thrust the engine propeller combination will need to produce to keep up the speed. The math is simple: this costs fuel.

We as the builder can try to keep the weight as low as possible (more lift results in more induced drag) but by paying attention while we construct the airframe we can also reduce parasite drag by streamlining parts as much as possible.

And by reducing the amount of drag (parasite) and keeping the empty weight to a minimum we consequently reduce fuel consumption and we can increase the payload and/or the range of our aircraft. We therefore have increased the efficiency of the aircraft/engine combination and the amount of money we can spent flying.

## Some Aerodynamics

For an aircraft in unaccelerated level flight, the thrust is equal to the drag of the aircraft, remember Newton's first law. And we also know that Thrust (Drag) = Horsepower × propeller efficiency. Both variables thrust and drag are expressed in either pounds or kilograms.

This formula tells us that if you increase the thrust or decrease the drag, or even do both, the aircraft will go faster. But there is a mathematical relationship between power, drag and speed. The speed increases with the cube root of the power increase or in formula form: Speed increase = Power increase$^{1/3}$.

Take this example to clarify: your aircraft has a IAS of 100 kts on a 150 bhp Lycoming engine and you want to know how fast it will fly with a 180 bhp Lycoming. The increase in power is 1,2 (results from 180/150). The cube root of that increase ($1,2^{1/3}$) equals 1,063. Thus the increase in speed is 100 × 1,063 = 106,3 Kts. In reality you will not go that much faster by increasing engine power alone.

## Higher Speeds

Drag also increases with the square of the speed. So if your aircraft total drag was 200 lbs at 100 Kts then increasing the speed to 106,3 Kts the total drag would end up to be $(106,3 / 100)^2 \times 200$ lbs = 226 lbs. That totals to 13 % more drag from a 6,3 % increase in speed.

We can conclude that drag reduction is the single best way to get additional speed from a certain design, and the faster the airplane, the more important each little improvement becomes.

# Lift-induced Drag

Induced Drag is an inevitable consequence of lift and is produced by the passage of an aerofoil (e.g. wing or tailplane) through the air. Air flowing over the top of a wing tends to flow inwards because the decreased pressure over the top surface is less than the pressure outside the wing tip. Below the wing, the air flows outwards because the pressure below the wing is greater than that outside the wing tip. The direct consequence of this, as far as the wing tips are concerned, is that there is a continual spilling of air upwards around the wing tip a phenomenon called 'tip effect' or 'end effect'. One way to appreciate why a high aspect ratio for a wing is better than a low one is that with a high aspect ratio, the proportion of air which moves in this way is reduced and therefore more of it generates lift.

For the wing more generally, the streams of air from above and below the wing are flowing at an angle to each other as they meet along the trailing edge of the wing. They combine to form vortices which, when viewed from the rear, rotate clockwise from the left wing and counter clockwise from the right. The tendency is for these vortices to move outwards towards the wing tip joining together as they do so. By the time the wing tip is reached, one large wig tip vortex has formed and is shed.

Most of these vortices are of course completely invisible but, in very humid air, the central core of a vortex may become visible because the air pressure within its centre has reduced - and has therefore cooled - sufficiently for condensation to occur. A higher wing loading in a turn will also increase the strength - and the degree of reduced pressure - so that visible vortex cores are even more likely during turns. If close up to these vortices.

Most of the air flowing off the top of a wing - 'downwash' - continues more or less horizontally towards the empennage because it is balanced by a corresponding upwash in front of the wing leading edge. In contrast, the upwards air movement which leads to vortex 'consolidation' at the tip is just outside the tip whereas the corresponding downward movement is just at the extremity of the wingspan so that the net direction of airflow past the wing is downwards. The lift created by the wing - which is by definition at right angles to the airflow, is therefore inclined slightly backwards and thus 'contributes' drag - induced drag.

## Source of Induced Drag

The total aerodynamic force acting on a body is usually thought of as having two components, lift and drag. By definition, the component of force parallel to the oncoming flow is called drag; and the component perpendicular to the oncoming flow is called lift. At practical angles of attack the lift greatly exceeds the drag.

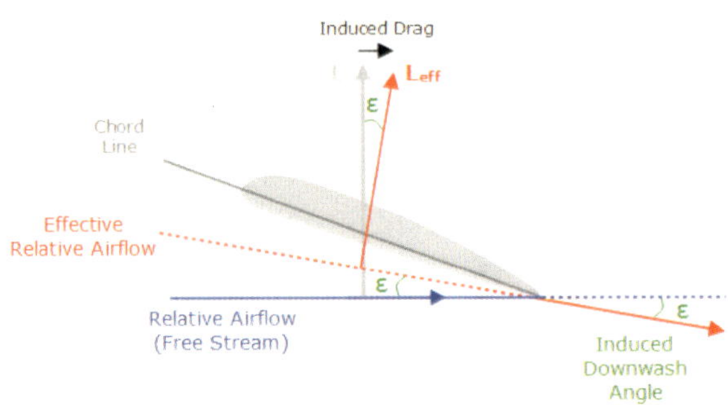

Induced drag is related to the amount of induced downwash in the vicinity of the wing.

The grey vertical line labeled "L" is perpendicular to the free stream and indicates the orientation of the lift on the wing. The red vector labeled "$L_{eff}$" is perpendicular to the actual airflow in the vicinity of the wing; it represents the lift on the airfoil section in two–dimensional flow at the same angle of attack. The lift generated by the wing has been tilted rearwards through an angle equal to the angle of the downwash in three-dimensional flow. The component of "$L_{eff}$" parallel to the free stream is the induced drag on the wing.

Lift is produced by the changing direction of the flow around a wing. The change of direction results in a change of velocity (even if there is no speed change, just as seen in uniform circular motion), which is an acceleration. To change the direction of the flow therefore requires that a force be applied to the fluid; lift is simply the reaction force of the fluid acting on the wing.

To produce lift, air below the wing is at a higher pressure than the air pressure above the wing. On a wing of finite span, this pressure difference causes air to flow from the lower surface wing root, around the wingtip, towards the upper surface wing root. This spanwise flow of air combines with chordwise flowing air, causing a change in speed and direction, which twists the airflow and produces vortices along the wing trailing edge. The vortices created are unstable, and they quickly combine to produce wingtip vortices. The resulting vortices change the speed and direction of the airflow behind the trailing edge, deflecting it downwards, and thus inducing downwash behind the wing.

Wingtip vortices modify the airflow around a wing, reducing wing's ability to generate lift, so that it requires a higher angle of attack for the same lift, which tilts the total aerodynamic force rearwards and increases the drag component of that force. The angular deflection is small and has little effect on the lift. However, there is an increase in the drag equal to the product of the lift force and the angle through which it is deflected. Since the deflection is itself a function of the lift, the additional drag is proportional to the square of the lift.

## Reducing Induced Drag

According to the equations below, a wing of infinite aspect ratio (wingspan/chord length) and constant airfoil section would produce no induced drag. The characteristics of such a wing can be measured on a section of wing spanning the width of a wind tunnel, since the walls block spanwise flow and create what is effectively two-dimensional flow.

A rectangular planform wing produces stronger wingtip vortices than does a tapered or elliptical wing, therefore many modern wings are tapered. However, an elliptical planform is more efficient as the induced downwash (and therefore the effective angle of attack) is constant across the whole of the wingspan. Few aircraft have this planform because of manufacturing complications — the most famous examples being the World War II Spitfire and Thunderbolt. Tapered wings with straight leading and trailing edges can approximate to elliptical lift distribution. Typically, straight edged non-tapered wings produce 5%, and tapered wings produce 1-2% more induced drag than an elliptical wing.

Similarly, for a given wing area, a high aspect ratio wing will produce less induced drag than a wing of low aspect ratio because there is less air disturbance at the tip of a longer, thinner wing. Induced drag can therefore be said to be inversely proportional to aspect ratio. The lift distribution may also be modified by the use of washout, a spanwise twist of the wing to reduce the incidence towards the wingtips, and by changing the airfoil section near the wingtips. This allows more lift to be generated nearer the wing root and less towards the wingtip, which causes a reduction in the strength of the wingtip vortices.

Some early aircraft had fins mounted on the tips of the tailplane which served as endplates. More recent aircraft have wingtip mounted winglets to reduce the intensity of wingtip vortices. Wingtip mounted fuel tanks may also provide some benefit, by preventing the spanwise flow of air around the wingtip.

## Calculation of Induced Drag

For a *planar* wing with an elliptical lift distribution, induced drag can be calculated as follows:

$$D_i = \frac{L^2}{\frac{1}{2}\rho V^2 \pi b^2},$$

where

$L$ is the lift,

$\rho$ is the air density,

$V$ is the true airspeed, and

$b$ is the wingspan.

From this equation it is clear that the induced drag decreases with flight speed and with wingspan. Deviation from the non-planar wing with elliptical lift distribution are taken into account by dividing the induced drag by the span efficiency factor $e$.

To compare with other sources of drag, it can be convenient to express this equation in terms of lift and drag coefficients:

$$C_{D,i} = \frac{D_i}{\frac{1}{2}\rho V^2 S} = \frac{C_L^2}{\pi e AR}, \text{ where}$$

$$C_L = \frac{L}{\frac{1}{2}\rho V^2 S}$$

and

$AR = \frac{b^2}{S}$ is the aspect ratio,

$S$ is a reference wing area.

This indicates how high aspect ratio wings are beneficial to flight efficiency. With $C_L$ being a function of angle of attack, induced drag increases as the angle of attack increases.

The above equation can be derived using Prandtl's lifting-line theory. Similar methods can also be used to compute the minimum induced drag for non-planar wings or for arbitrary lift distributions.

### Combined Effect with other Drag Sources

Induced drag must be added to the parasitic drag to find the total drag. Since induced drag is inversely proportional to the square of the airspeed (at a given lift) whereas parasitic drag is proportional to the square of the airspeed, the combined overall drag curve shows a minimum at some airspeed - the minimum drag speed ($V_{MD}$). An aircraft flying at this speed is operating at its optimal aerodynamic efficiency. According to the above equations, the speed for minimum drag occurs at the speed where the induced drag is equal to the parasitic drag. This is the speed at which for unpowered aircraft, optimum glide angle is achieved. This is also the speed for greatest range (although $V_{MD}$ will decrease as the plane consumes fuel and becomes lighter). The speed for greatest range (i.e., distance travelled) is the speed at which a straight line from the origin is tangent to the fuel flow rate curve. The curve of range versus airspeed is normally very flat and it is customary to operate at the speed for 99% best range since this gives about 5% greater speed for only 1% less range. (Of course, flying higher where the air is thinner will raise the speed at which minimum drag occurs, and so permits a faster voyage for the same amount of fuel. If the plane is flying at the maximum permissible speed, then there is an altitude at which the air density will be what is needed to keep it aloft while flying at the angle of attack that minimizes the drag. The optimum altitude at maximum speed, and the optimum speed at maximum altitude, may change during the flight as the plane becomes lighter).

The speed for maximum endurance (i.e., time in the air) is the speed for minimum fuel flow rate, and is less than the speed for greatest range. The fuel flow rate is calculated as the product of the power required and the engine specific fuel consumption (fuel flow rate per unit of power). The power required is equal to the drag times the speed.

## Center of Pressure

As an object moves through a fluid, the velocity of the fluid varies around the surface of the object. The variation of velocity produces a variation of pressure on the surface of the object as shown by the the thin red lines on the figure. Integrating the pressure times the surface area around the body determines the aerodynamic force on the object. We can consider this single force to act through the average location of the pressure on the surface of the object. We call the average location of the pressure variation the center of pressure in the same way that we call the average location of the weight of an object the center of gravity. The aerodynamic force can then be resolved into two components, lift and drag, which act through the center of pressure in flight.

Determining the center of pressure is very important for any flying object. To trim an airplane, or to provide stability for a model rocket or a kite, it is necessary to know the location of the center of pressure of the entire aircraft. How do engineers determine the location of the center of pressure for an aircraft which they are designing?

In general, determining the center of pressure (cp) is a very complicated procedure because the pressure changes around the object. Determining the center of pressure requires the use of calculus and a knowledge of the pressure distribution around the body. We can characterize the pressure variation around the surface as a function p(x) which indicates that the pressure depends on the distance x from a reference line usually taken as the leading edge of the object. If we can determine the form of the function, there are methods to perform a calculus integration of the equation. We will use the symbols "S[ ]dx" to denote the integration of a continuous function. Then the center of pressure can be determined from:

cp = (S[x * p(x)]dx) / (S[p(x)]dx)

If we don't know the actual functional form, we can numerically integrate the equation using a spreadsheet by dividing the distance into a number of small distance segments and determining the average value of the pressure over that small segment. Taking the sum of the average value times the distance times the distance segment divided by the sum of the average value times the distance segment will produce the center of pressure.

## Aircraft Aerodynamics

A stable configuration is desirable not only in sailing, but in aircraft design as well. Aircraft design therefore borrowed the term center of pressure. And like a sail, a rigid non-symmetrical airfoil not only produces lift, but a moment. The center of pressure of an aircraft is the point where all of the aerodynamic pressure field may be represented by a single force vector with no moment. A similar idea is the aerodynamic center which is the point on an airfoil where the pitching moment produced by the aerodynamic forces is constant with angle of attack.

The aerodynamic center plays an important role in analysis of the longitudinal static stability of all flying machines. It is desirable that when the pitch angle and angle of attack of an aircraft are disturbed (by, for example turbulence) that the aircraft returns to its original trimmed pitch angle and angle of attack without a pilot or autopilot changing the control surface deflection. For an aircraft to return towards its trimmed attitude, without input from a pilot or autopilot, it must have positive longitudinal static stability.

## Missile Aerodynamics

Missiles typically do not have a preferred plane or direction of maneuver and thus have symmetric airfoils. Since the center of pressure for symmetric airfoils is relatively constant for small angle of attack, missile engineers typically speak of the complete center of pressure of the entire vehicle for stability and control analysis. In missile analysis, the center of pressure is typically defined as the center of the additional pressure field due to a change in the angle of attack off of the trim angle of attack.

For unguided rockets the trim position is typically zero angle of attack and the center of pressure is defined to be the center of pressure of the resultant flow field on the entire vehicle resulting from a very small angle of attack (that is, the center of pressure in the limit as angle of attack goes to zero). For positive stability in missiles, the total vehicle center of pressure defined as given above must be further from the nose of the vehicle than the center of gravity. In missiles at lower angles of attack, the contributions to the center of pressure are dominated by the nose, wings, and fins. The normalized normal force coefficient derivative with respect to the angle of attack of each component multiplied by the location of the center of pressure can be used to compute a centroid representing the total center of pressure. The center of pressure of the added flow field is behind the center of gravity and the additional force "points" in the direction of the added angle of attack; this produces a moment that pushes the vehicle back to the trim position.

In guided missiles where the fins can be moved to trim the vehicles in different angles of attack, the center of pressure is the center of pressure of the flow field at that angle of attack for the undeflected fin position. This is the center of pressure of any small change in the angle of attack (as defined above). Once again for positive static stability, this definition of center of pressure requires that the center of pressure be further from the nose than the center of gravity. This ensures that any increased forces resulting from increased angle of attack results in increased restoring moment to drive the missile back to the trimmed position. In missile analysis, positive static margin implies that the complete vehicle makes a restoring moment for any angle of attack from the trim position.

## Movement of Center of Pressure for Aerodynamic Fields

The center of pressure on a symmetric airfoil typically lies close to 25% of the chord length behind the leading edge of the airfoil. (This is called the "quarter-chord point"). For a symmetric airfoil, as angle of attack and lift coefficient change, the center of pressure does not move. It remains around the quarter-chord point for angles of attack below the stalling angle of attack. The role of center of pressure in the control characterization of aircraft takes a different form than in missiles.

On a cambered airfoil the center of pressure does not occupy a fixed location. For a conventionally cambered airfoil, the center of pressure lies a little behind the quarter-chord point at maximum lift coefficient (large angle of attack), but as lift coefficient reduces (angle of attack reduces) the center of pressure moves toward the rear. When the lift coefficient is zero an airfoil is generating no lift but a conventionally cambered airfoil generates a nose-down pitching moment, so the location of the center of pressure is an infinite distance behind the airfoil.

For a reflex-cambered airfoil, the center of pressure lies a little ahead of the quarter-chord point at maximum lift coefficient (large angle of attack), but as lift coefficient reduces (angle of attack reduces) the center of pressure moves forward. When the lift coefficient is zero an airfoil is generating no lift but a reflex-cambered airfoil generates a nose-up pitching moment, so the location of the center of pressure is an infinite distance ahead of the airfoil. This direction of movement of the center of pressure on a reflex-cambered airfoil has a stabilising effect.

The way the center of pressure moves as lift coefficient changes makes it difficult to use the center of pressure in the mathematical analysis of longitudinal static stability of an aircraft. For this reason, it is much simpler to use the aerodynamic center when carrying out a mathematical analysis. The aerodynamic center occupies a fixed location on an airfoil, typically close to the quarter-chord point.

The aerodynamic center is the conceptual starting point for longitudinal stability. The horizontal stabilizer contributes extra stability and this allows the center of gravity to be a small distance aft of the aerodynamic center without the aircraft reaching neutral stability. The position of the center of gravity at which the aircraft has neutral stability is called the neutral point.

# Thrust

Thrust is the force which moves an aircraft through the air. Thrust is used to overcome the drag of an airplane, and to overcome the weight of a rocket. Thrust is generated by the engines of the aircraft through some kind of propulsion system.

Thrust is a mechanical force, so the propulsion system must be in physical contact with a working fluid to produce thrust. Thrust is generated most often through the reaction of accelerating a mass of gas. Since thrust is a force, it is a vector quantity having both a magnitude and a direction. The engine does work on the gas and accelerates the gas to the rear of the engine; the thrust is generated in the opposite direction from the accelerated gas. The magnitude of the thrust depends on the amount of gas that is accelerated and on the difference in velocity of the gas through the engine.

To accelerate the gas, we have to expend energy. The energy is generated as heat by the combustion of some fuel. The thrust equation describes how the acceleration of the gas produces a force. The type of propulsion system used on an aircraft may vary from airplane to airplane and each device produces thrust in a slightly different way.

## Examples

A fixed-wing aircraft generates forward thrust when air is pushed in the direction opposite to flight. This can be done in several ways including by the spinning blades of a propeller, or a rotating fan pushing air out from the back of a jet engine, or by ejecting hot gases from a rocket engine. The forward thrust is proportional to the mass of the airstream multiplied by the difference in velocity of the airstream. Reverse thrust can be generated to aid braking after landing by reversing the pitch of variable-pitch propeller blades, or using a thrust reverser on a jet engine. Rotary wing aircraft and thrust vectoring V/STOL aircraft use engine thrust to support the weight of the aircraft, and vector sum of this thrust fore and aft to control forward speed.

A rocket is propelled forward by a thrust force equal in magnitude, but opposite in direction, to the time-rate of momentum change of the exhaust gas accelerated from the combustion chamber through the rocket engine nozzle. This is the exhaust velocity with respect to the rocket, times the time-rate at which the mass is expelled, or in mathematical terms:

$$\mathrm{T} = \mathrm{v}\frac{dm}{dt}$$

Where T is the thrust generated (force), $\frac{dm}{dt}$ is the rate of change of mass with respect to time (mass flow rate of exhaust), and v is the speed of the exhaust gases measured relative to the rocket. For vertical launch of a rocket the initial thrust at liftoff must be more than the weight.

Each of the three Space Shuttle Main Engines could produce a thrust of 1.8 MN, and each of the Space Shuttle's two Solid Rocket Boosters 14.7 MN, together 29.4 MN.

By contrast, the simplified Aid For EVA Rescue (SAFER) has 24 thrusters of 3.56 N each.

In the air-breathing category, the AMT-USA AT-180 jet engine developed for radio-controlled aircraft produce 90 N (20 lbf) of thrust. The GE90-115B engine fitted on the Boeing 777-300ER, recognized by the Guinness Book of World Records as the "World's Most Powerful Commercial Jet Engine," has a thrust of 569 kN (127,900 lbf).

## Concepts

### Thrust to Power

The power needed to generate thrust and the force of the thrust can be related in a non-linear way. In general, $P^2 \propto T^3$. The proportionality constant varies, and can be solved for a uniform flow:

$$\frac{dm}{dt} = \rho A v$$

$$T = \frac{dm}{dt} v, P = \frac{1}{2} \frac{dm}{dt} v^2$$

$$T = \rho A v^2, P = \frac{1}{2} \rho A v^3$$

$$P^2 = \frac{T^3}{4 \rho A}$$

Note that these calculations are only valid for when the incoming air is accelerated from a standstill - for example when hovering.

The inverse of the proportionality constant, the "efficiency" of an otherwise-perfect thruster, is proportional to the area of the cross section of the propelled volume of fluid ( $A$ ) and the density of the fluid ( $\rho$ ). This helps to explain why moving through water is easier and why aircraft have much larger propellers than watercraft.

### Thrust to Propulsive Power

A very common question is how to contrast the thrust rating of a jet engine with the power rating of a piston engine. Such comparison is difficult, as these quantities are not equivalent. A piston engine does not move the aircraft by itself (the propeller does that), so piston engines are usually rated by how much power they deliver to the propeller. Except for changes in temperature and air pressure, this quantity depends basically on the throttle setting.

A jet engine has no propeller, so the propulsive power of a jet engine is determined from its thrust as follows. Power is the force (F) it takes to move something over some distance (d) divided by the time (t) it takes to move that distance:

$$P = F \frac{d}{t}$$

In case of a rocket or a jet aircraft, the force is exactly the thrust (T) produced by the engine. If the rocket or aircraft is moving at about a constant speed, then distance divided by time is just speed, so power is thrust times speed:

$$P = Tv$$

This formula looks very surprising, but it is correct: the *propulsive power* (or *power available*) of a jet engine increases with its speed. If the speed is zero, then the propulsive power is zero. If a jet aircraft is at full throttle but attached to a static test stand, then the jet engine produces no propulsive power, however thrust is still produced. Compare that to a piston engine. The combination piston engine–propeller also has a propulsive power with exactly the same formula, and it will also be zero at zero speed -- but that is for the engine–propeller set. The engine alone will continue to produce its rated power at a constant rate, whether the aircraft is moving or not.

Now, imagine the strong chain is broken, and the jet and the piston aircraft start to move. At low speeds:

Thepistonenginewillhaveconstant100%power,andthepropeller'sthrustwillvarywithspeed
The jet engine will have constant 100% thrust, and the engine's power will vary with speed.

### Excess Thrust

If a powered aircraft is generating thrust T and experiencing drag D, the difference between the two, T – D, is termed the excess thrust. The instantaneous performance of the aircraft is mostly dependent on the excess thrust.

Excess thrust is a vector and is determined as the vector difference between the thrust vector and the drag vector.

### Centre of Thrust

The centre of thrust for an object is an average point at which the total thrust may be considered to apply. It may differ from the centre of gravity.

## Boundary Layer

As an object moves through a fluid, or as a fluid moves past an object, the molecules of the fluid near the object are disturbed and move around the object. Aerodynamic forces are generated between the fluid and the object. The magnitude of these forces depend on the shape of the object, the speed of the object, the mass of the fluid going by the object and on two other important properties of the fluid; the viscosity, or stickiness, and the compressibility, or springiness, of the fluid. To properly model these effects, aerospace engineers use similarity parameters which are ratios of these effects to other forces present in the problem. If two experiments have the same values for the similarity parameters, then the relative importance of the forces are being correctly modeled.

Aerodynamic forces depend in a complex way on the viscosity of the fluid. As the fluid moves past the object, the molecules right next to the surface stick to the surface. The molecules just above the surface are slowed down in their collisions with the molecules sticking to the surface. These molecules in turn slow down the flow just above them. The farther one moves away from the surface, the fewer the collisions affected by the object surface. This creates a thin layer of fluid near the surface in which the velocity changes from zero at the surface to the free stream value away from the surface. Engineers call this layer the boundary layer because it occurs on the boundary of the fluid.

The details of the flow within the boundary layer are very important for many problems in aerodynamics, including wing stall, the skin friction drag on an object, and the heat transfer that occurs in high speed flight.

## Types of Boundary Layer

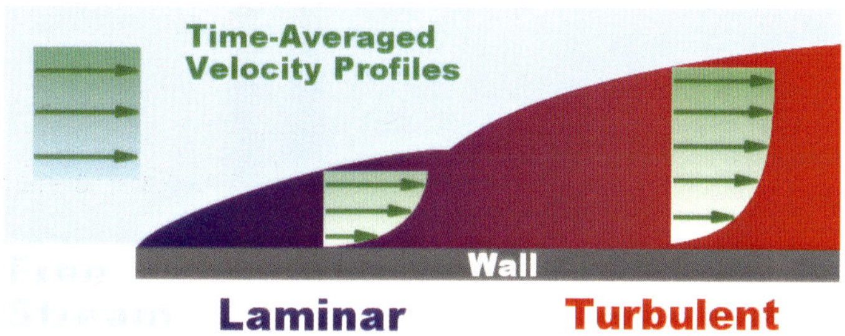


Boundary layer visualization, showing transition from laminar to turbulent condition

Laminar boundary layers can be loosely classified according to their structure and the circumstances under which they are created. The thin shear layer which develops on an oscillating body is an example of a Stokes boundary layer, while the Blasius boundary layer refers to the well-known similarity solution near an attached flat plate held in an oncoming unidirectional flow and Falkner–Skan boundary layer, a generalization of Blasius profile. When a fluid rotates and viscous forces are balanced by the Coriolis effect (rather than convective inertia), an Ekman layer forms. In the theory of heat transfer, a thermal boundary layer occurs. A surface can have multiple types of boundary layer simultaneously.

The viscous nature of airflow reduces the local velocities on a surface and is responsible for skin friction. The layer of air over the wing's surface that is slowed down or stopped by viscosity, is the boundary layer. There are two different types of boundary layer flow: laminar and turbulent.

## Laminar Boundary Layer Flow

The laminar boundary is a very smooth flow, while the turbulent boundary layer contains swirls or "eddies." The laminar flow creates less skin friction drag than the turbulent flow, but is less stable. Boundary layer flow over a wing surface begins as a smooth laminar flow. As the flow continues back from the leading edge, the laminar boundary layer increases in thickness.

## Turbulent Boundary Layer Flow

At some distance back from the leading edge, the smooth laminar flow breaks down and transitions to a turbulent flow. From a drag standpoint, it is advisable to have the transition from laminar to turbulent flow as far aft on the wing as possible, or have a large amount of the wing surface within the laminar portion of the boundary layer. The low energy laminar flow, however, tends to break down more suddenly than the turbulent layer.

## Aerodynamics

Ludwig Prandtl

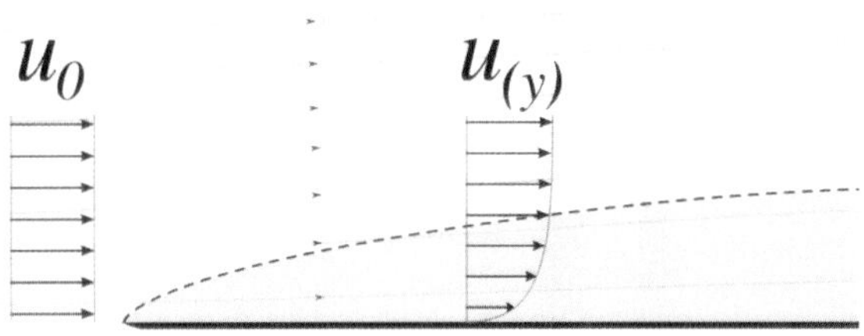


Laminar boundary layer velocity profile

The aerodynamic boundary layer was first defined by Ludwig Prandtl in a paper presented on August 12, 1904 at the third International Congress of Mathematicians in Heidelberg, Germany. It simplifies the equations of fluid flow by dividing the flow field into two areas: one inside the boundary layer, dominated by viscosity and creating the majority of drag experienced by the boundary body; and one outside the boundary layer, where viscosity can be neglected without significant effects on the solution. This allows a closed-form solution for the flow in both areas, a significant simplification of the full Navier–Stokes equations. The majority of the heat transfer to and from a body also takes place within the boundary layer, again allowing the equations to be simplified in the flow field outside the boundary layer. The pressure distribution throughout the boundary layer in the direction normal to the surface (such as an airfoil) remains constant throughout the boundary layer, and is the same as on the surface itself.

The thickness of the velocity boundary layer is normally defined as the distance from the solid body to the point at which the viscous flow velocity is 99% of the freestream velocity (the surface velocity of an inviscid flow). Displacement thickness is an alternative definition stating that the boundary layer represents a deficit in mass flow compared to inviscid flow with slip at the wall. It is the distance by which the wall would have to be displaced in the inviscid case to give the same total mass flow as the viscous case. The no-slip condition requires the flow velocity at the surface of a solid object be zero and the fluid temperature be equal to the temperature of the surface. The flow velocity will then increase rapidly within the boundary layer, governed by the boundary layer equations.

The thermal boundary layer thickness is similarly the distance from the body at which the temperature is 99% of the freestream temperature. The ratio of the two thicknesses is governed by the Prandtl number. If the Prandtl number is 1, the two boundary layers are the same thickness. If the Prandtl number is greater than 1, the thermal boundary layer is thinner than the velocity boundary layer. If the Prandtl number is less than 1, which is the case for air at standard conditions, the thermal boundary layer is thicker than the velocity boundary layer.

In high-performance designs, such as gliders and commercial aircraft, much attention is paid to controlling the behavior of the boundary layer to minimize drag. Two effects have to be considered. First, the boundary layer adds to the effective thickness of the body, through the displacement thickness, hence increasing the pressure drag. Secondly, the shear forces at the surface of the wing create skin friction drag.

At high Reynolds numbers, typical of full-sized aircraft, it is desirable to have a laminar boundary layer. This results in a lower skin friction due to the characteristic velocity profile of laminar flow. However, the boundary layer inevitably thickens and becomes less stable as the flow develops along the body, and eventually becomes turbulent, the process known as boundary layer transition. One way of dealing with this problem is to suck the boundary layer away through a porous surface. This can reduce drag, but is usually impractical due to its mechanical complexity and the power required to move the air and dispose of it. Natural laminar flow techniques push the boundary layer transition aft by reshaping the aerofoil or fuselage so that its thickest point is more aft and less thick. This reduces the velocities in the leading part and the same Reynolds number is achieved with a greater length.

At lower Reynolds numbers, such as those seen with model aircraft, it is relatively easy to maintain laminar flow. This gives low skin friction, which is desirable. However, the same velocity profile which gives the laminar boundary layer its low skin friction also causes it to be badly affected by adverse pressure gradients. As the pressure begins to recover over the rear part of the wing chord, a laminar boundary layer will tend to separate from the surface. Such flow separation causes a large increase in the pressure drag, since it greatly increases the effective size of the wing section. In these cases, it can be advantageous to deliberately trip the boundary layer into turbulence at a point prior to the location of laminar separation, using a turbulator. The fuller velocity profile of the turbulent boundary layer allows it to sustain the adverse pressure gradient without separating. Thus, although the skin friction is increased, overall drag is decreased. This is the principle behind the dimpling on golf balls, as well as vortex generators on aircraft. Special wing sections have also been designed which tailor the pressure recovery so laminar separation is reduced or even eliminated. This represents an optimum compromise between the pressure drag from flow separation and skin friction from induced turbulence.

When using half-models in wind tunnels, a peniche is sometimes used to reduce or eliminate the effect of the boundary layer.

## Boundary Layer Equations

The deduction of the boundary layer equations was one of the most important advances in fluid dynamics. Using an order of magnitude analysis, the well-known governing Navier–Stokes equations of viscous fluid flow can be greatly simplified within the boundary layer. Notably, the characteristic of the partial differential equations (PDE) becomes parabolic, rather than the elliptical form of the full Navier–Stokes equations. This greatly simplifies the solution of the equations. By making the boundary layer approximation, the flow is divided into an inviscid portion (which is easy to solve by a number of methods) and the boundary layer, which is governed by an easier to solve PDE. The continuity and Navier–Stokes equations for a two-dimensional steady incompressible flow in Cartesian coordinates are given by

$$\frac{\partial u}{\partial x}+\frac{\partial \upsilon}{\partial y}=0$$

$$u\frac{\partial u}{\partial x}+\upsilon\frac{\partial u}{\partial y}=-\frac{1}{\rho}\frac{\partial p}{\partial x}+\nu\left(\frac{\partial^2 u}{\partial x^2}+\frac{\partial^2 u}{\partial y^2}\right)$$

$$u\frac{\partial \upsilon}{\partial x}+\upsilon\frac{\partial \upsilon}{\partial y}=-\frac{1}{\rho}\frac{\partial p}{\partial y}+\nu\left(\frac{\partial^2 \upsilon}{\partial x^2}+\frac{\partial^2 \upsilon}{\partial y^2}\right)$$

where $u$ and $\upsilon$ are the velocity components, $\rho$ is the density, $p$ is the pressure, and $\nu$ is the kinematic viscosity of the fluid at a point.

The approximation states that, for a sufficiently high Reynolds number the flow over a surface can be divided into an outer region of inviscid flow unaffected by viscosity (the majority of the flow), and a region close to the surface where viscosity is important (the boundary layer). Let $u$ and $\upsilon$ be streamwise and transverse (wall normal) velocities respectively inside the boundary layer. Using scale analysis, it can be shown that the above equations of motion reduce within the boundary layer to become.

$$u\frac{\partial u}{\partial x}+\upsilon\frac{\partial u}{\partial y}=-\frac{1}{\rho}\frac{\partial p}{\partial x}+\nu\frac{\partial^2 u}{\partial y^2}$$

$$\frac{1}{\rho}\frac{\partial p}{\partial y}=0$$

and if the fluid is incompressible (as liquids are under standard conditions):

$$\frac{\partial u}{\partial x}+\frac{\partial \upsilon}{\partial y}=0$$

The order of magnitude analysis assumes the streamwise length scale significantly larger than the transverse length scale inside the boundary layer. It follows that variations in properties in the streamwise direction are generally much lower than those in the wall normal direction. Apply this to the continuity equation shows that $\upsilon$, the wall normal velocity, is small compared with $u$ the streamwise velocity.

Since the static pressure $p$ is independent of $y$, then pressure at the edge of the boundary layer is the pressure throughout the boundary layer at a given streamwise position. The external pressure may be obtained through an application of Bernoulli's equation. Let $U$ be the fluid velocity outside the boundary layer, where $u$ and $U$ are both parallel. This gives upon substituting for $p$ the following result

$$u\frac{\partial u}{\partial x}+\upsilon\frac{\partial u}{\partial y}=U\frac{dU}{dx}+\nu\frac{\partial^2 u}{\partial y^2}$$

For a flow in which the static pressure $p$ also does not change in the direction of the flow

$$\frac{dp}{dx}=0$$

so $U$ remains constant.

Therefore, the equation of motion simplifies to become

$$u\frac{\partial u}{\partial x}+\upsilon\frac{\partial u}{\partial y}=\nu\frac{\partial^2 u}{\partial y^2}$$

These approximations are used in a variety of practical flow problems of scientific and engineering interest. The above analysis is for any instantaneous laminar or turbulent boundary layer, but is used mainly in laminar flow studies since the mean flow is also the instantaneous flow because there are no velocity fluctuations present. This simplified equation is a parabolic PDE and can be solved using a similarity solution often referred to as the Blasius boundary layer.

## Prandtl's Transposition Theorem

Prandtl observed that from any solution $u(x,y,t), v(x,y,t)$ which satisfies the boundary layer equations, further solution $u^*(x,y,t), v^*(x,y,t)$, which is also satisfying the boundary layer equations, can be constructed by writing

$$u^*(x,y,t)=u(x,y+f(x),t),\quad v^*(x,y,t)=v(x,y+f(x),t)-f'(x)u(x,y+f(x),t)$$

where $f(x)$ is arbitrary. Since the solution is not unique from mathematical perspective, to the solution can added any one of an infinite set of eigenfunctions as shown by Stewartson and Paul A. Libby.

## Von Kármán Momentum Integral

Von Kármán derived the integral equation by integrating the boundary layer equation across the boundary layer in 1921. The equation is

$$\frac{\tau_w}{\rho U^2}=\frac{1}{U^2}\frac{\partial}{\partial t}(U\delta_1)+\frac{\partial \delta_2}{\partial x}+\frac{2\delta_2+\delta_1}{U}\frac{\partial U}{\partial x}+\frac{v_w}{U}$$

where

$$\tau_w = \mu\left(\frac{\partial u}{\partial y}\right)_{y=0}, \quad v_w = v(x,0,t), \quad \delta_1 = \int_0^\infty \left(1-\frac{u}{U}\right)dy, \quad \delta_2 = \int_0^\infty \frac{u}{U}\left(1-\frac{u}{U}\right)dy$$

$\tau_w$ is the wall shear stress, $v_w$ is the suction/injection velocity at the wall, $\delta_1$ is the displacement thickness and $\delta_2$ is the momentum thickness. Kármán-Pohlhausen Approximation is derived from this equation.

### Energy Integral

The energy integral was derived by Wieghardt.

$$\frac{2\varepsilon}{\rho U^3} = \frac{1}{U}\frac{\partial}{\partial t}(\delta_1 + \delta_2) + \frac{2\delta_2}{U^2}\frac{\partial U}{\partial t} + \frac{1}{U^3}\frac{\partial}{\partial x}(U^3\delta_3) + \frac{v_w}{U}$$

where

$$\varepsilon = \int_0^\infty \mu\left(\frac{\partial u}{\partial y}\right)^2 dy, \quad \delta_3 = \int_0^\infty \frac{u}{U}\left(1-\frac{u^2}{U^2}\right)dy$$

$\varepsilon$ is the energy dissipation rate due to viscosity across the boundary layer and $\delta_3$ is the energy thickness.

### Von Mises Transformation

For steady two-dimensional boundary layers, von Mises introduced a transformation which takes $x$ and $\psi$ (stream function) as independent variables instead of $x$ and $y$ and uses a dependent variable $\chi = U^2 - u^2$ instead of $u$. The boundary layer equation then become

$$\frac{\partial \chi}{\partial x} = \nu\sqrt{U^2 - \chi}\,\frac{\partial^2 \chi}{\partial \psi^2}$$

The original variables are recovered from

$$y = \int \sqrt{U^2 - \chi}\,d\psi, \quad u = \sqrt{U^2 - \chi}, \quad v = u\int \frac{\partial}{\partial x}\left(\frac{1}{u}\right)d\psi.$$

This transformation is later extended to compressible boundary layer by von Kármán and HS Tsien.

### Crocco's Transformation

For steady two-dimensional compressible boundary layer, Luigi Crocco introduced a transformation which takes $x$ and $u$ as independent variables instead of $x$ and $y$ and uses a dependent variable $\tau = \mu\partial u/\partial y$ (shear stress) instead of $u$. The boundary layer equation then become

$$\mu\rho u\frac{\partial}{\partial x}\left(\frac{1}{\tau}\right) + \frac{\partial^2 \tau}{\partial u^2} - \mu\frac{dp}{dx}\frac{\partial}{\partial u}\left(\frac{1}{\tau}\right) = 0$$

$$\text{if } \frac{dp}{dx} = 0, \quad \frac{\mu\rho}{\tau^2}\frac{\partial \tau}{\partial x} = \frac{1}{u}\frac{\partial^2 \tau}{\partial u^2}$$

The original coordinate is recovered from

$$y = \mu \int \frac{du}{\tau}.$$

## Turbulent Boundary Layers

The treatment of turbulent boundary layers is far more difficult due to the time-dependent variation of the flow properties. One of the most widely used techniques in which turbulent flows are tackled is to apply Reynolds decomposition. Here the instantaneous flow properties are decomposed into a mean and fluctuating component. Applying this technique to the boundary layer equations gives the full turbulent boundary layer equations not often given in literature:

$$\frac{\partial \overline{u}}{\partial x} + \frac{\partial \overline{v}}{\partial y} =$$

$$\overline{u}\frac{\partial \overline{u}}{\partial x} + \overline{v}\frac{\partial \overline{u}}{\partial y} = -\frac{1}{\rho}\frac{\partial \overline{p}}{\partial x} + \nu\left(\frac{\partial^2 \overline{u}}{\partial x^2} + \frac{\partial^2 \overline{u}}{\partial y^2}\right) - \frac{\partial}{\partial y}(\overline{u'v'}) - \frac{\partial}{\partial x}(\overline{u'^2})$$

$$\overline{u}\frac{\partial \overline{v}}{\partial x} + \overline{v}\frac{\partial \overline{v}}{\partial y} = -\frac{1}{\rho}\frac{\partial \overline{p}}{\partial y} + \nu\left(\frac{\partial^2 \overline{v}}{\partial x^2} + \frac{\partial^2 \overline{v}}{\partial y^2}\right) - \frac{\partial}{\partial x}(\overline{u'v'}) - \frac{\partial}{\partial y}(\overline{v'^2})$$

Using a similar order-of-magnitude analysis, the above equations can be reduced to leading order terms. By choosing length scales $\delta$ for changes in the transverse-direction, and $L$ for changes in the streamwise-direction, with $\delta << L$, the x-momentum equation simplifies to:

$$\overline{u}\frac{\partial \overline{u}}{\partial x} + \overline{v}\frac{\partial \overline{u}}{\partial y} = -\frac{1}{\rho}\frac{\partial \overline{p}}{\partial x} - \frac{\partial}{\partial y}(\overline{u'v'}).$$

This equation does not satisfy the no-slip condition at the wall. Like Prandtl did for his boundary layer equations, a new, smaller length scale must be used to allow the viscous term to become leading order in the momentum equation. By choosing $\eta << \delta$ as the y-scale, the leading order momentum equation for this "inner boundary layer" is given by:

$$0 = -\frac{1}{\rho}\frac{\partial \overline{p}}{\partial x} + \nu\frac{\partial^2 \overline{u}}{\partial y^2} - \frac{\partial}{\partial y}(\overline{u'v'}).$$

In the limit of infinite Reynolds number, the pressure gradient term can be shown to have no effect on the inner region of the turbulent boundary layer. The new "inner length scale" $\eta$ is a viscous length scale, and is of order $\frac{\nu}{u_*}$, with $u_*$ being the velocity scale of the turbulent fluctuations, in this case a friction velocity.

Unlike the laminar boundary layer equations, the presence of two regimes governed by different sets of flow scales (i.e. the inner and outer scaling) has made finding a universal similarity solution for the turbulent boundary layer difficult and controversial. To find a similarity solution that spans both regions of the flow, it is necessary to asymptotically match the solutions from both regions of the flow. Such analysis will yield either the so-called log-law or power-law.

The additional term $\overline{u'v'}$ in the turbulent boundary layer equations is known as the Reynolds shear stress and is unknown a priori. The solution of the turbulent boundary layer equations therefore necessitates the use of a turbulence model, which aims to express the Reynolds shear stress in terms of known flow variables or derivatives. The lack of accuracy and generality of such models is a major obstacle in the successful prediction of turbulent flow properties in modern fluid dynamics.

A constant stress layer exists in the near wall region. Due to the damping of the vertical velocity fluctuations near the wall, the Reynolds stress term will become negligible and we find that a linear velocity profile exists. This is only true for the very near wall region.

## Compressible Flow

As an aircraft moves through the air, the air molecules near the aircraft are disturbed and move around the aircraft. If the aircraft passes at a low speed, typically less than 250 mph, it is observed that the density of the air remains constant. For higher speeds, some of the energy of the aircraft goes into compressing the air and locally changing the density of the air. This compressibility effect alters the amount of resulting force on the aircraft since the aerodynamic force depends on the air density. The effect becomes more important as speed increases. Near and beyond the speed of sound, about 330 m/s or 760 mph at sea level, small disturbances in the flow are transmitted to other locations isentropically or with constant entropy. Sharp disturbances generate shock waves that affect both the lift and drag of the aircraft, and the flow conditions downstream of the shock wave.

The Mach number is the ratio of the speed of the aircraft, or the speed of the gas, to the speed of sound in the gas. The speed of sound is equal to the speed of transmission of small, isentropic disturbances in the flow. To determine the role of the Mach number on compressibility effects. we begin with the conservation of momentum equation:

rho * V dV = - dp

where rho is the fluid density, V is the velocity, and p is the pressure. dV and dp denote differential changes in the velocity and pressure. From our derivation of the conditions for isentropic flow, we know that:

dp/p = gamma * drho/rho

dp = gamma * p / rho * drho

where gamma is the specific heat ratio. We can use the ideal equation of state to simplify the expression on the right:

p = rho * R * T

dp = gamma * R * T * drho

where R is the specific gas constant and T is the absolute temperature. We recognize that:

gamma * R * T = a^2

where a is equal to the speed of sound. So,

dp = a^2 * drho

Substituting this expression for the change of pressure into the conservation of momentum equation gives:

rho * V dV = - a^2 drho

- (V^2 / a^2) dV / V = drho / rho

- M^2 dV / V = drho / rho

where M is the Mach number.

What does this expression tell us about the role of the Mach number in compressible flows?

1. For low speed, or subsonic conditions, the Mach number is less than one, M < 1 and the square of the Mach number is very small. Then the left hand side of the equation is very small, and the change in density is very small. For the low subsonic conditions, compressibility can be ignored.

2. As the speed of the object approaches the speed of sound, the flight Mach number is nearly equal to one, M = 1, and the flow is said to be transonic. If the Mach number is near one, the square of the Mach number is also nearly equal to one. For transonic flows, the change in density is nearly equal to the change in velocity, and compressibility effects can not be ignored.

3. As the speed increases beyond the speed of sound, the flight Mach number is greater than one M > 1 and the flow is said to be supersonic or hypersonic. For supersonic and hypersonic flows, the density changes faster than the velocity changes by a factor equl to the square of the Mach number. Compressibility effects become more important with higher Mach numbers.

## Introductory Concepts

There are several important assumptions involved in the underlying theory of compressible flow. All fluids are composed of molecules, but tracking a huge number of individual molecules in a flow (for example at atmospheric pressure) is unnecessary. Instead, the continuum assumption allows us to consider a flowing gas as a continuous substance except at low densities. This assumption provides a huge simplification which is accurate for most gas-dynamic problems. Only in the low-density realm of rarefied gas dynamics does the motion of individual molecules become important.

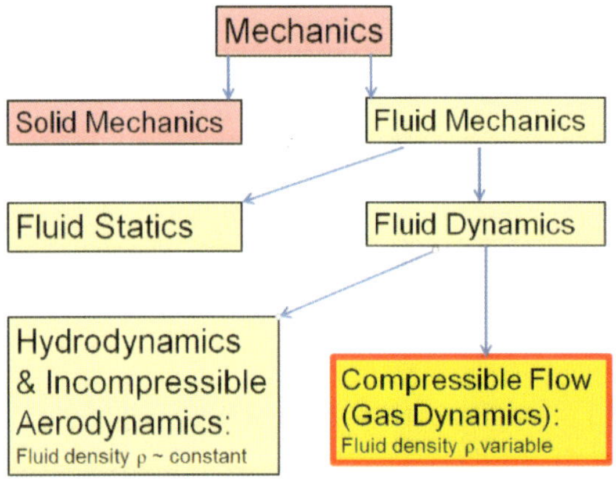


Breakdown of fluid mechanics chart

A related assumption is the no-slip condition where the flow velocity at a solid surface is presumed equal to the velocity of the surface itself, which is a direct consequence of assuming continuum flow. The no-slip condition implies that the flow is viscous, and as a result a boundary layer forms on bodies traveling through the air at high speeds, much as it does in low-speed flow.

Most problems in incompressible flow involve only two unknowns: pressure and velocity, which are typically found by solving the two equations that describe conservation of mass and of linear momentum, with the fluid density presumed constant. In compressible flow, however, the gas density and temperature also become variables. This requires two more equations in order to solve compressible-flow problems: an equation of state for the gas and a conservation of energy equation. For the majority of gas-dynamic problems, the simple Ideal gas law is the appropriate state equation.

Fluid dynamics problems have two overall types of references frames, called Lagrangian and Eulerian. The Lagrangian approach follows a fluid mass of fixed identity as it moves through a flowfield. The Eulerian reference frame, in contrast, does not move with the fluid. Rather it is a fixed frame or control volume that fluid flows through. The Eulerian frame is most useful in a majority of compressible flow problems, but requires that the equations of motion be written in a compatible format.

Finally, although space is known to have 3 dimensions, an important simplification can be had in describing gas dynamics mathematically if only one spatial dimension is of primary importance, hence 1-dimensional flow is assumed. This works well in duct, nozzle, and diffuser flows where the flow properties change mainly in the flow direction rather than perpendicular to the flow. However, an important class of compressible flows, including the external flow over bodies traveling at high speed, requires at least a 2-dimensional treatment. When all 3 spatial dimensions and perhaps the time dimension as well are important, we often resort to computerized solutions of the governing equations.

## One-dimensional Flow

One-dimensional (1-D) flow refers to the flow of gas through a duct or channel in which the flow parameters are assumed to change significantly along only one spatial dimension, namely, the duct length. In analysing the 1-D channel flow, a number of assumptions are made:

- Ratio of duct length to width (L/D) is ≤ about 5 (in order to neglect friction and heat transfer),
- Steady vs. Unsteady Flow,

- Flow is isentropic (i.e. a reversible adiabatic process),
- Ideal gas law (i.e. $P=\rho RT$).

## Converging-diverging Laval Nozzles

As the speed of a flow accelerates from the subsonic to the supersonic regime, the physics of nozzle and diffuser flows is altered. Using the conservation laws of fluid dynamics and thermodynamics, the following relationship for channel flow is developed (combined mass and momentum conservation):

$$dP(1-M^2)=\rho V^2\left(\frac{dA}{A}\right),$$

where dP is the differential change in pressure, M is the Mach number, ρ is the density of the gas, V is the velocity of the flow, A is the area of the duct, and dA is the change in area of the duct. This equation states that, for subsonic flow, a converging duct ($dA<0$) increases the velocity of the flow and a diverging duct ($dA>0$) decreases velocity of the flow. For supersonic flow, the opposite occurs due to the change of sign of ($1-M^2$). A converging duct ($dA<0$) now decreases the velocity of the flow and a diverging duct ($dA>0$) increases the velocity of the flow. At Mach = 1, a special case occurs in which the duct area must be either a maximum or minimum. For practical purposes, only a minimum area can accelerate flows to Mach 1 and beyond. See table of sub-supersonic diffusers and nozzles.

| | Subsonic M<1 | Supersonic M>1 |
|---|---|---|
| Converging Duct 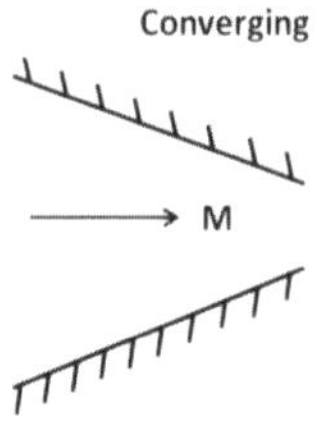 | As P decreases, then *v* increases<br>Subsonic nozzle | As P increases, then *v* decreases<br>Supersonic diffuser |
| Diverging Duct 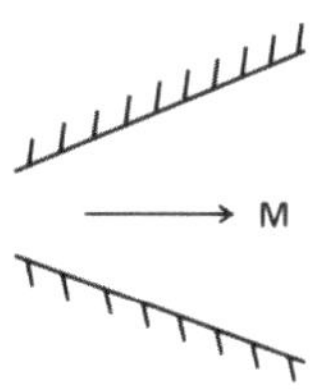 | As P increases, then *v* decreases<br>Subsonic diffuser | As P decreases, then *v* increases<br>Supersonic nozzle |

Table showing the reversal in the physics of nozzles and diffusers with changing Mach numbers

Therefore, to accelerate a flow to Mach 1, a nozzle must be designed to converge to a minimum cross-sectional area and then expand. This type of nozzle – the converging-diverging nozzle – is called a de Laval nozzle after Gustaf de Laval, who invented it. As subsonic flow enters the converging duct and the area decreases, the flow accelerates. Upon reaching the minimum area of the duct, also known as the throat of the nozzle, the flow can reach Mach 1. If the speed of the flow is to continue to increase, its density must decrease in order to obey conservation of mass. To achieve this decrease in density, the flow must expand, and to do so, the flow must pass through a diverging duct.

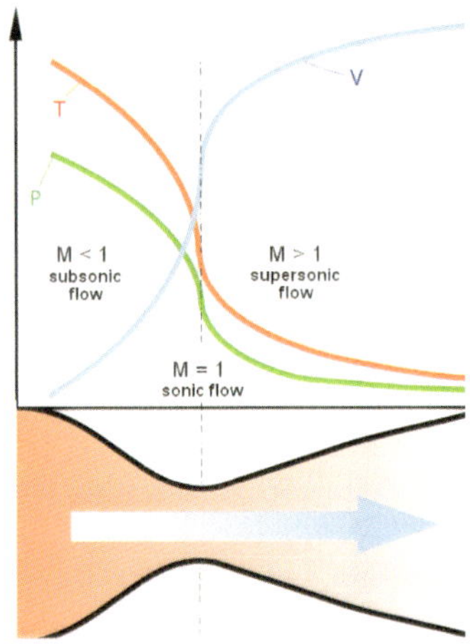


Nozzle de Laval diagram

## Maximum Achievable Velocity of a Gas

Ultimately, because of the energy conservation law, a gas is limited to a certain maximum velocity based on its energy content. The maximum velocity, $V_{max}$, that a gas can attain is:

$$V_{max} = \sqrt{2c_p T_t}$$

where $c_p$ is the specific heat of the gas and $T_t$ is the stagnation temperature of the flow.

## Isentropic Flow Mach Number Relationships

Using conservations laws and thermodynamics, a number of relationships of the form

$$\frac{property_1}{property_2} = f(M, \gamma)$$

can be obtained, where M is the Mach number and γ is the ratio of specific heats (1.4 for air). See table of isentropic flow Mach number relationships.

| Isentropic Flow Mach Number Relationships: | |
|---|---|
| Stagnation to Static Temperature | $\frac{T_t}{T} = \left(1 + \frac{\gamma - 1}{2} M^2\right)$ |
| Stagnation to Static Pressure | $\frac{P_t}{P} = \left(1 + \frac{\gamma - 1}{2} M^2\right)^{\frac{\gamma}{\gamma - 1}}$ |
| Mass Flow Rate | $\dot{m} = \frac{P_t}{\sqrt{RT_t}} A\sqrt{\gamma}\, M \left(1 + \frac{\gamma - 1}{2} M^2\right)^{\frac{\gamma+1}{2-2\gamma}}$ |
| Stagnation to Static Density | $\frac{\rho_t}{\rho} = \left(1 + \frac{\gamma - 1}{2} M^2\right)^{\frac{1}{\gamma - 1}}$ |
| Arbitrary Area to Choked Area in Duct | $\frac{A}{A*} = \frac{1}{M}\left(\frac{\frac{\gamma + 1}{2}}{1 + \frac{\gamma - 1}{2} M^2}\right)^{\frac{\gamma+1}{2-2\gamma}}$ |

Isentropic flow relationship table. Equations to relate the field properties in isentropic flow.

## Achieving Supersonic Flow

In order for a flow to become supersonic, it must pass through a duct with a minimum area, or sonic throat. Additionally, an overall pressure ratio, $P_b/P_t$, of approximately 2 is needed to attain Mach 1. Once it has reached Mach 1, the flow at the throat is said to be "choked." Because changes downstream can only move upstream at sonic speed, the mass flow through the nozzle cannot be affected by changes in downstream conditions after the flow is choked.

## Non-isentropic 1D Channel Flow of a Gas - Normal Shock Waves

Normal shock waves are shock waves that are perpendicular to the local flow direction. These shock waves occur when pressure waves build up and coalesce into an extremely thin shock-wave that converts useful energy into heat. The waves thus overtake and reinforce one another, forming a finite shock wave from an infinite series of infinitesimal sound waves. Because a loss of energy occurs over the thin shock wave, the shock is considered non-isentropic and enthalpy increases across the shock. When analysing a normal shock wave, one-dimensional, steady, and adiabatic (stagnation temperature does not change across the shock wave) flow of a perfect gas is assumed.

| **Rankine-Hugoniot Equations:** | |
|---|---|
| **Temperature** | $\frac{T_2}{T_1} = \frac{\left(1+\frac{\gamma-1}{2}M_1^2\right)\left(\frac{2\gamma}{\gamma-1}M_1^2-1\right)}{M_1^2\left(\frac{2\gamma}{\gamma-1}+\frac{\gamma-1}{2}\right)}$ |
| **Static Pressure** | $\frac{P_2}{P_1} = \frac{2\gamma M_1^2}{\gamma+1} - \frac{\gamma-1}{\gamma+1}$ |
| **Density and Velocity** | $\frac{\rho_2}{\rho_1} = \frac{v_2}{v_1} = \frac{(\gamma+1)M_1^2}{(\gamma-1)M_1^2+2}$ |

The Rankine-Hugoniot equations relate conditions before and after a normal shock wave.

Normal shock waves can occur in two reference frames: the standing normal shock and the moving shock. The flow before a normal shock wave must be supersonic, and the flow after a normal shock must be subsonic. The Rankine-Hugoniot equations are used to solve for the flow conditions.

## Two-dimensional Flow

Although one-dimensional flow can be directly analysed, it is merely a specialized case of two-dimensional flow. It follows that one of the defining phenomena of one-dimensional flow, a normal shock, is likewise only a special case of a larger class of oblique shocks. Further, the name "normal" is with respect to geometry rather than frequency of occurrence. Oblique shocks are much more common in applications such as: aircraft inlet design, objects in supersonic flight, and (at a more fundamental level) supersonic nozzles and diffusers. Depending on the flow conditions, an oblique shock can either be attached to the flow or detached from the flow in the form of a bow shock.

Attached shock wave shown on a X-15 Model in a supersonic wind tunnel

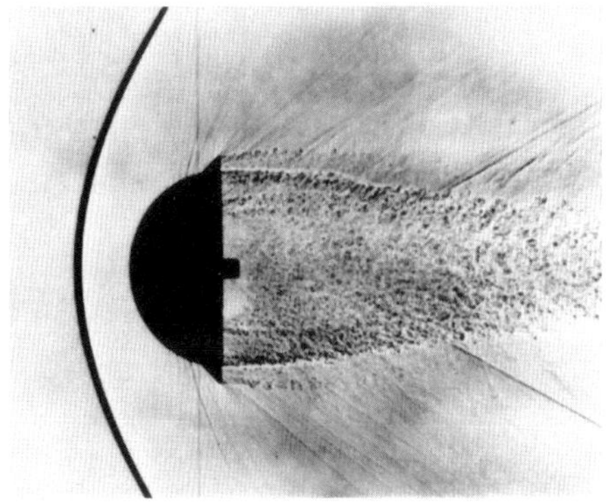

Bowshock example for a blunt body

## Oblique Shock Waves

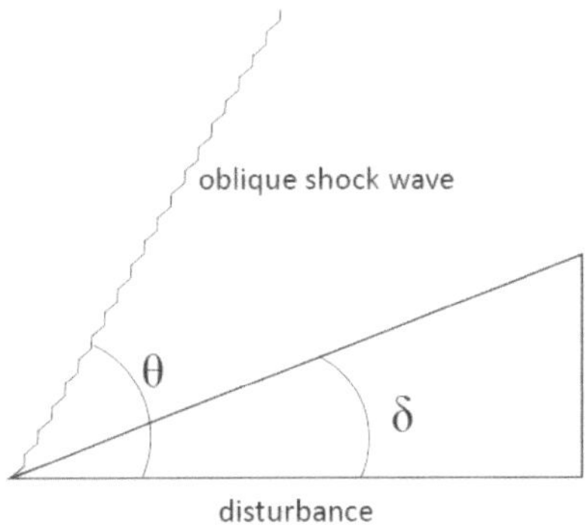


Diagram of obstruction

Oblique shock waves are similar to normal shock waves, but they occur at angles less than 90° with the direction of flow. When a disturbance is introduced to the flow at a nonzero angle (δ), the flow must respond to the changing boundary conditions. Thus an oblique shock is formed, resulting in a change in the direction of the flow.

## Shock Polar Diagram

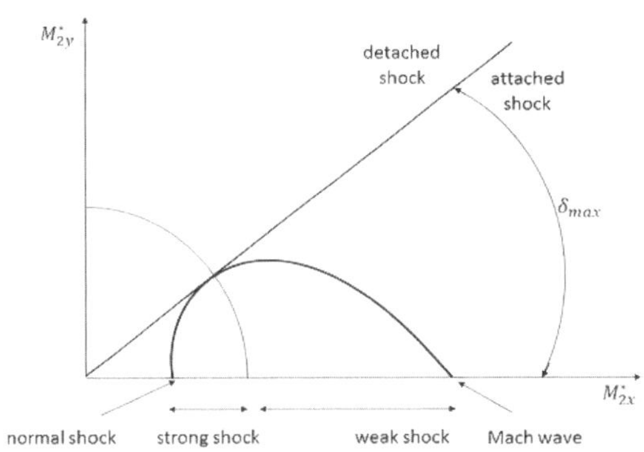


Shock polar diagram

Based on the level of flow deflection (δ), oblique shocks are characterized as either strong or weak. Strong shocks are characterized by larger deflection and more entropy loss across the shock, with weak shocks as the opposite. In order to gain cursory insight into the differences in these shocks, a shock polar diagram can be used. With the static temperature after the shock, T*, known the speed of sound after the shock is defined as,

$$A^* = \sqrt{\gamma R T^*}$$

with R as the gas constant and γ as the specific heat ratio. The Mach number can be broken into Cartesian coordinates

$$M_{2x}^* = \frac{V_x}{a^*}$$

$$M_{2y}^* = \frac{V_y}{a^*}$$

with $V_x$ and $V_y$ as the x and y-components of the fluid velocity V. With the Mach number before the shock given, a locus of conditions can be specified. At some $\delta_{max}$ the flow transitions from a strong to weak oblique shock. With δ = 0°, a normal shock is produced at the limit of the strong oblique shock and the Mach wave is produced at the limit of the weak shock wave.

## Oblique Shock Reflection

Due to the inclination of the shock, after an oblique shock is created, it can interact with a boundary in three different manners.

## Solid Boundary

Incoming flow is first turned by angle δ with respect to the flow. This shockwave is reflected off the solid boundary, and the flow is turned by – δ to again be parallel with the boundary. It is important to note that each progressive shock wave is weaker and the wave angle is increased.

## Irregular Reflection

An irregular reflection is much like the case described above, with the caveat that δ is larger than the maximum allowable turning angle. Thus a detached shock is formed and a more complicated reflection occurs.

## Prandtl-meyer Fans

Prandtl-Meyer fans can be expressed as both compression and expansion fans. Prandtl-Meyer fans also cross a boundary layer (i.e. flowing and solid) which reacts in different changes as well. When a shock wave hits a solid surface the resulting fan returns as one from the opposite family while when one hits a free boundary the fan returns as a fan of opposite type.

## Prandtl-meyer Expansion Fans

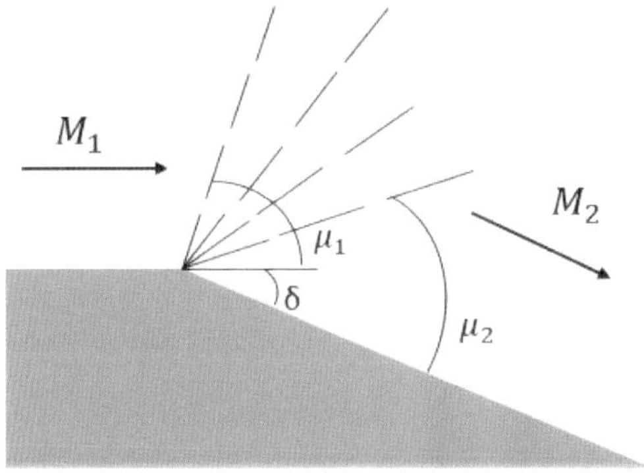

Prandtl-Meyer expansion fan diagram

To this point, the only flow phenomena that have been discussed are shock waves, which slow the flow and increase its entropy. It is possible to accelerate supersonic flow in what has been termed a Prandtl-Meyer expansion fan, after Ludwig Prandtl and Theodore Meyer.

As opposed to the flow encountering an inclined obstruction and forming an oblique shock, the flow expands around a convex corner and forms an expansion fan through a series of isentropic Mach waves. The expansion "fan" is composed of Mach waves that span from the initial Mach angle to the final Mach angle. Flow can expand around either a sharp or rounded corner equally, as the increase in Mach number is proportional to only the convex angle of the passage (δ). The expansion corner that produces the Prandtl-Meyer fan can be sharp (as illustrated in the figure) or rounded. If the total turning angle is the same, then the P-M flow solution is also the same.

The Prandtl-Meyer expansion can be seen as the physical explanation of the operation of the Laval nozzle. The contour of the nozzle creates a smooth and continuous series of Prandtl-Meyer expansion waves.

## Prandtl-meyer Compression Fans

A Prandtl-Meyer compression is the opposite phenomenon to a Prandtl-Meyer expansion. If the flow is gradually turned through an angle of δ, a compression fan can be formed. This fan is a series of Mach waves that eventually coalesce into an oblique shock. Because the flow is defined by an isentropic region (flow that travels through the fan) and an anisentropic region (flow that travels through the oblique shock), a slip line results between the two flow regions.

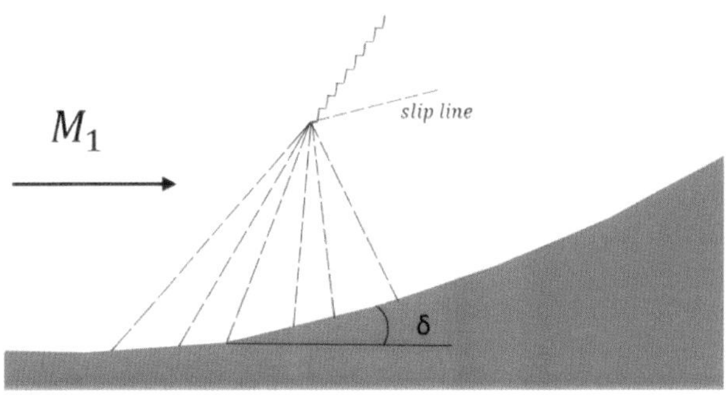

Basic PM compression diagram

# Applications

## Supersonic Wind Tunnels

Supersonic wind tunnels are used for testing and research in supersonic flows, approximately over the Mach number range of 1.2 to 5. The operating principle behind the wind tunnel is that a large pressure difference is maintained upstream to downstream, driving the flow.

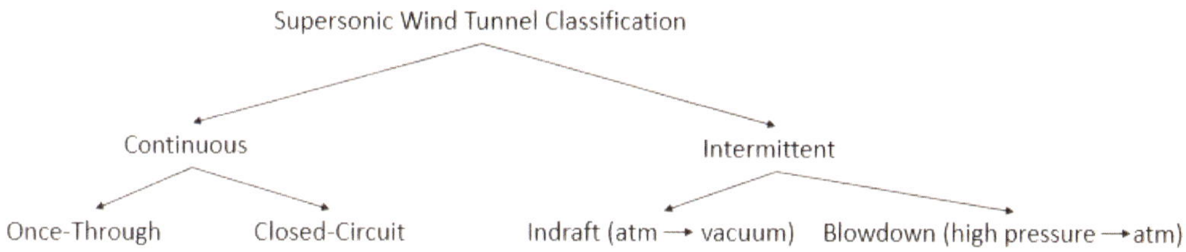


Supersonic wind tunnel classification list

Wind tunnels can be divided into two categories: continuous-operating and intermittent-operating wind tunnels. Continuous operating supersonic wind tunnels require an independent electrical power source that drastically increases with the size of the test section. Intermittent supersonic wind tunnels are less expensive in that they store electrical energy over an extended period of time, then discharge the energy over a series of brief tests. The difference between these two is analogous to the comparison between a battery and a capacitor.

Blowdown type supersonic wind tunnels offer high Reynolds number, a small storage tank, and readily available dry air. However, they cause a high pressure hazard, result in difficulty holding a constant stagnation pressure, and are noisy during operation.

Indraft supersonic wind tunnels are not associated with a pressure hazard, allow a constant stagnation pressure, and are relatively quiet. Unfortunately, they have a limited range for the Reynolds number of the flow and require a large vacuum tank. There is no dispute that knowledge is gained through research and testing in supersonic wind tunnels; however, the facilities often require vast amounts of power to maintain the large pressure ratios needed for testing conditions. For example, Arnold Engineering Development Complex has the largest supersonic wind tunnel in the world and requires the power required to light a small city for operation. For this reason, large wind tunnels are becoming less common at universities.

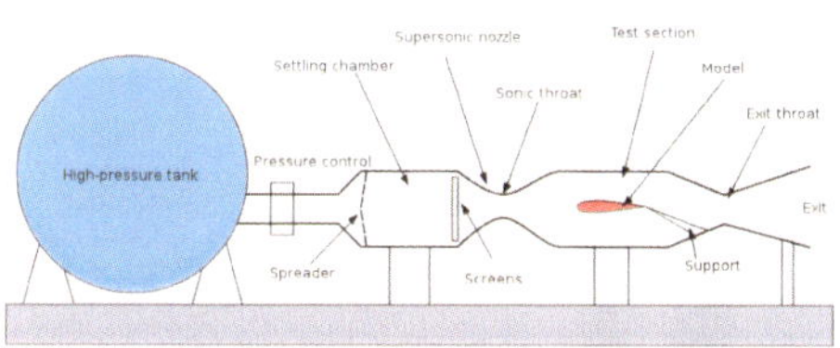


Blowdown supersonic wind tunnel schematic

Langley indraft supersonic wind tunnel vacuum sphere

## Supersonic Aircraft Inlets

Perhaps the most common application for oblique shocks is in high-speed aircraft inlets. The purpose of the inlet is to slow incoming supersonic flow to the subsonic regime before it enters the turbojet engine, with the caveat of minimizing losses across the shock. Knowledge of normal and oblique shocks suggests that this be accomplished with a series of weakening oblique shocks followed by a very weak normal shock, usually less than M = 1.4. This may sound relatively straightforward, but there is one rather large issue to be dealt with when designing a supersonic aircraft inlet: acceleration. Between taking off, manoeuvring, and cruising, an aircraft travels at a range of Mach numbers. In order to ensure efficient flight, the aircraft intake must be capable of variable geometry. If it is not, the shock waves will not reflect properly through the inlet and negatively affect performance.

McDonnell Douglas F-15C

Concorde

Lockheed SR-71 Blackbird

Although variable geometry is a universally recognized approach to improve aircraft efficiency and performance over a range of Mach numbers, there is no one method to achieve variable geometry. The F-15 Eagle employs wedge inlets with adjustable flaps to control the flow. For subsonic flow, the flaps are completely closed and for supersonic flow, the flaps are open. The Concorde employed an external-compression inlet, using a series of oblique shocks followed by a normal shock to slow the flow sufficiently for the turbojet engine. Perhaps the most recognizable supersonic aircraft, the SR-71, used a hydraulically actuated cone to reduce the speed of the supersonic flow through the aircraft inlet.

# Transonic

As an aircraft moves through the air, the air molecules near the aircraft are disturbed and move around the aircraft. Exactly how the air re-acts to the aircraft depends upon the ratio of the speed of the aircraft to the speed of sound through the air. Because of the importance of this speed ratio, aerodynamicists have designated it with a special parameter called the Mach number in honor of Ernst Mach, a late 19th century physicist who studied gas dynamics.

In aeronautics, transonic refers to the condition of flight in which a range of velocities of airflow exist surrounding and flowing past an air vehicle or an airfoil that are concurrently below, at, and above the speed of sound in the range of Mach 0.8 to 1.2. This condition depends not only on the travel speed of the craft, but also on the temperature of the airflow in the vehicle's local environment. It is formally defined as the range of speeds between the critical Mach number, when some parts of the airflow over an air vehicle or airfoil are supersonic, and a higher speed, typically near Mach 1.2, when the vast majority of the airflow is supersonic. Between these speeds some of the airflow is supersonic, but a significant fraction is not.

The first powered aircraft to explore this regime were the high performance fighters of World War II. These aircraft seemed to encounter a sound barrier at which drag was increasing faster than thrust. There was speculation in the mid-1940's that manned flight was not possible at speeds faster than the speed of sound, even though the muzzle velocity of rifle bullets is supersonic. Of course, the flight of the X-1A in 1947 proved that people could fly faster than sound and, until the recent retirement of the Concorde, any person with enough money can fly supersonic. As mentioned above, even though modern airliners typically fly at about M = .85, the flow over the wings is transonic or supersonic. Drag increases dramatically as an aircraft approaches Mach 1, so airliners use high thrust gas turbine propulsion systems. On the slide we show a DC-8 airliner which is powered by four turbofan engines. The wings of airliners are typically swept in planform to reduce the transonic drag. For Mach numbers less than 2.0, the frictional heating of the airframe is low enough that light weight aluminum is used for the structure.

Most modern jet powered aircraft are engineered to operate at transonic air speeds. Transonic airspeeds see a rapid increase of drag from about Mach 0.8, and it is the fuel costs of that drag that typically limits the airspeed. Attempts to reduce wave drag can be seen on all high-speed aircraft; most notable is the use of swept wings, but another common form is a wasp-waist fuselage as a side effect of the Whitcomb area rule.

Transonic cloud forms around shock wave as aircraft passes through the sound barrier.

Severe instability can occur at transonic speeds. Shock waves move through the air at the speed of sound. When an object such as an aircraft also moves at the speed of sound, these shock waves build up in front of it to form a single, very large shock wave. During transonic flight, the plane must pass through this large shock wave, as well as contend with the instability caused by air moving faster than sound over parts of the wing and slower in other parts.

Transonic speeds can also occur at the tips of rotor blades of helicopters and aircraft. However, as this puts severe, unequal stresses on the rotor blade and may lead to dangerous accidents if it occurs, it is avoided. Blade tip speed is one of the limiting factors to the size of rotors, and also to the forward speeds of helicopters (as this speed is added to the forward-sweeping (leading) side of the rotor, thus possibly causing localized transonics).

## Condensation Clouds

At transonic speeds supersonic expansion fans form intense low-pressure, low-temperature areas at various points around an aircraft. If the temperature drops below the dew point a visible cloud will form. These clouds remain with the aircraft as it travels. It is not necessary for the aircraft as a whole to reach supersonic speeds for these clouds to form. Typically, the tail of the aircraft will reach supersonic flight while the bow of the aircraft is still in subsonic flight. A bubble of supersonic expansion fans terminating by a wake shockwave surround the tail. As the aircraft continues to accelerate, the supersonic expansion fans will intensify and the wake shockwave will grow in size until infinity is reached, at which point the bow shockwave forms. This is Mach one and the Prandtl–Glauert singularity.

## Mach Number

Mach number, in fluid mechanics, ratio of the velocity of a fluid to the velocity of sound in that fluid, named after Ernst Mach (1838–1916), an Austrian physicist and philosopher. In the case of an object moving through a fluid, such as an aircraft in flight, the Mach number is equal to the velocity of the object relative to the fluid divided by the velocity of sound in that fluid. Mach numbers less than one indicate subsonic flow; those greater than one, supersonic flow. Fluid flow, in addition, is classified as compressible or incompressible on the basis of the Mach number. For example, gas flowing with a Mach number of less than three-tenths may be considered incompressible, or of constant density, an approximation that greatly simplifies the analysis of its behaviour. For Mach numbers greater than one, shock wave patterns develop on the moving body because of compression of the surrounding fluid. Streamlining alleviates shock wave effects.

The Mach Number is a dimensionless value useful for analyzing fluid flow dynamics problems where compressibility is a significant factor.

The Mach Number can be expressed as

$M = v / c$

where

$M$ = Mach number

$v$ = fluid flow speed (m/s, ft/s)

$c$ = speed of sound (m/s, ft/s)

Alternatively the Mach Number can be expressed with the density and the bulk modulus for elasticity as

$M = v\,(\rho/E)^{1/2}$

where

$\rho$ = density of fluid (kg/m$^3$, lb/ft$^3$)

$E$ = bulk modulus elasticity (N/m$^2$ (Pa), lbf/in$^2$ (psi))

The bulk modulus elasticity has the dimension pressure and is commonly used to characterize the fluid compressibility.

The square of the Mach number is the Cauchy Number.

$M^2 = C$

where

$C$ = Cauchy Number

Mach number is useful because the fluid behaves in a similar manner at a given Mach number, regardless of other variables. As modeled in the International Standard Atmosphere, dry air at mean sea level, standard temperature of 15 °C (59 °F), the speed of sound is 340.3 meters per second (1,116.5 ft/s). The speed of sound is not a constant; in a gas, it increases as the absolute temperature increases, and since atmospheric temperature generally decreases with increasing altitude between sea level and 11,000 meters (36,089 ft), the speed of sound also decreases. For example, the standard atmosphere model lapses temperature to −56.5 °C (−69.7 °F) at 11,000 meters (36,089 ft) altitude, with a corresponding speed of sound (Mach 1) of 295.0 meters per second (967.8 ft/s), 86.7% of the sea level value.

## Classification of Mach Regimes

While the terms *subsonic* and *supersonic*, in the purest sense, refer to speeds below and above the local speed of sound respectively, aerodynamicists often use the same terms to talk about particular ranges of Mach values. This occurs because of the presence of a *transonic regime* around M = 1 where approximations of the Navier-Stokes equations used for subsonic design actually no longer apply; the simplest explanation is that the flow locally begins to exceed M = 1 even though the freestream Mach number is below this value.

Meanwhile, the *supersonic regime* is usually used to talk about the set of Mach numbers for which linearised theory may be used, where for example the (air) flow is not chemically reacting, and where heat-transfer between air and vehicle may be reasonably neglected in calculations.

In the following table, the *regimes* or *ranges of Mach values* are referred to, and not the *pure* meanings of the words *subsonic* and *supersonic*.

Generally, NASA defines *high* hypersonic as any Mach number from 10 to 25, and re-entry speeds as anything greater than Mach 25. Aircraft operating in this regime include the Space Shuttle and various space planes in development.

| Regime | Speed | | | | | General plane characteristics |
|---|---|---|---|---|---|---|
| | (Mach) | (knots) | (mph) | (km/h) | (m/s) | |
| Subsonic | <0.8 | <530 | <609 | <980 | <273 | Most often propeller-driven and commercial turbofan aircraft with high aspect-ratio (slender) wings, and rounded features like the nose and leading edges.<br><br>The subsonic speed range is that range of speeds within all of the airflow over an aircraft is less than Mach 1. The critical Mach number (Mcrit) is lowest free stream mach number at which airflow over any part of the aircraft first reaches Mach 1. So the subsonic speed range includes all speeds that are less than Mcrit. |

| | | | | | | |
|---|---|---|---|---|---|---|
| Transonic | 0.8–1.3 | 530–794 | 609–914 | 980 –1,470 | 273 –409 | Transonic aircraft nearly always have swept wings, causing the delay of drag-divergence, and often feature a design that adheres to the principles of the Whitcomb Area rule.<br><br>The transonic speed range is that range of speeds within which the airflow over different parts of an aircraft is between subsonic and supersonic. So the regime of flight from Mcrit up to Mach 1.3 is called the transonic range. |
| Supersonic | 1.3–5.0 | 794-3,308 | 915-3,806 | 1,470 –6,126 | 410 –1,702 | The supersonic speed range is that range of speeds within which all of the airflow over an aircraft is supersonic (more than Mach 1). But airflow meeting the leading edges is initially decelerated, so the free stream speed must be slightly greater than Mach 1 to ensure that all of the flow over the aircraft is supersonic. It is commonly accepted that the supersonic speed range starts at a free stream speed greater than Mach 1.3.<br><br>Aircraft designed to fly at supersonic speeds show large differences in their aerodynamic design because of the radical differences in the behaviour of flows above Mach 1. Sharp edges, thin aerofoil-sections, and all-moving tailplane/canards are common. Modern combat aircraft must compromise in order to maintain low-speed handling; "true" supersonic designs include the F-104 Starfighter, SR-71 Blackbird and BAC/Aérospatiale Concorde. |
| Hypersonic | 5.0 –10.0 | 3,308 –6,615 | 3,806 –7,680 | 6,126 –12,251 | 1,702 –3,403 | The X-15, at Mach 6.72 is one of the fastest manned aircraft. Also, cooled nickel-titanium skin; highly integrated (due to domination of interference effects: non-linear behaviour means that superposition of results for separate components is invalid), small wings, such as those on the Mach 5 X-51A Waverider. |
| High-hypersonic | 10.0 –25.0 | 6,615 –16,537 | 7,680 –19,031 | 12,251 –30,626 | 3,403 –8,508 | The NASA X-43, at Mach 9.6 is one of the fastest aircraft. Thermal control becomes a dominant design consideration. Structure must either be designed to operate hot, or be protected by special silicate tiles or similar. Chemically reacting flow can also cause corrosion of the vehicle's skin, with free-atomic oxygen featuring in very high-speed flows. Hypersonic designs are often forced into blunt configurations because of the aerodynamic heating rising with a reduced radius of curvature. |
| Re-entry speeds | >25.0 | >16,537 | >19,031 | >30,626 | >8,508 | Ablative heat shield; small or no wings; blunt shape |

## High-speed Flow Around Objects

Flight can be roughly classified in six categories:

| Regime | Subsonic | Transonic | Sonic | Supersonic | Hypersonic | Hypervelocity |
|---|---|---|---|---|---|---|
| Mach | <0.8 | 0.8–1.2 | 1.0 | 1.2–5.0 | 5.0–10.0 | >10.0 |

For comparison: the required speed for low Earth orbit is approximately 7.5 km/s = Mach 25.4 in air at high altitudes.

At transonic speeds, the flow field around the object includes both sub- and supersonic parts. The transonic period begins when first zones of $M > 1$ flow appear around the object. In case of an airfoil (such as an aircraft's wing), this typically happens above the wing. Supersonic flow can decelerate back to subsonic only in a normal shock; this typically happens before the trailing edge.

As the speed increases, the zone of $M > 1$ flow increases towards both leading and trailing edges. As $M = 1$ is reached and passed, the normal shock reaches the trailing edge and becomes a weak oblique shock: the flow decelerates over the shock, but remains supersonic. A normal shock is created ahead of the object, and the only subsonic zone in the flow field is a small area around the object's leading edge.

When an aircraft exceeds Mach 1 (i.e. the sound barrier), a large pressure difference is created just in front of the aircraft. This abrupt pressure difference, called a shock wave, spreads backward and outward from the aircraft in a cone shape (a so-called Mach cone). It is this shock wave that causes the sonic boom heard as a fast moving aircraft travels overhead. A person inside the aircraft will not hear this. The higher the speed, the more narrow the cone; at just over $M = 1$ it is hardly a cone at all, but closer to a slightly concave plane.

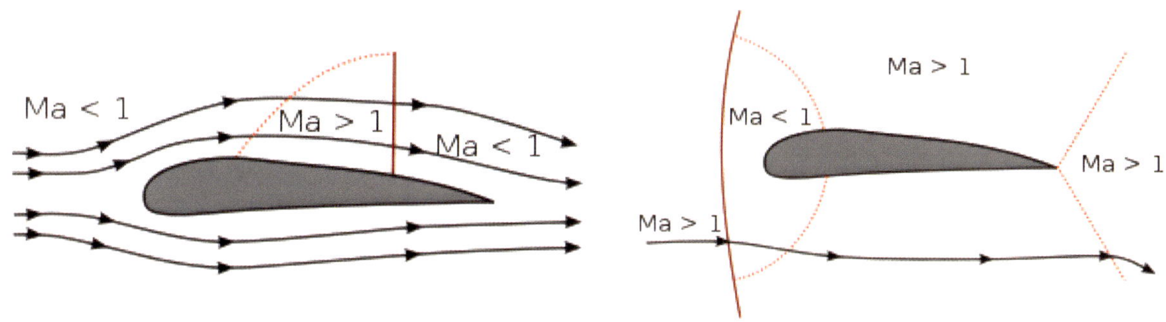


(a) (b)

*Mach number in transonic airflow around an airfoil; $M < 1$ (a) and $M > 1$ (b).*

At fully supersonic speed, the shock wave starts to take its cone shape and flow is either completely supersonic, or (in case of a blunt object), only a very small subsonic flow area remains between the object's nose and the shock wave it creates ahead of itself. (In the case of a sharp object, there is no air between the nose and the shock wave: the shock wave starts from the nose).

As the Mach number increases, so does the strength of the shock wave and the Mach cone becomes increasingly narrow. As the fluid flow crosses the shock wave, its speed is reduced and temperature, pressure, and density increase. The stronger the shock, the greater the changes. At high enough Mach numbers the temperature increases so much over the shock that ionization and dissociation of gas molecules behind the shock wave begin. Such flows are called hypersonic.

It is clear that any object traveling at hypersonic speeds will likewise be exposed to the same extreme temperatures as the gas behind the nose shock wave, and hence choice of heat-resistant materials becomes important.

## High-speed Flow in a Channel

As a flow in a channel becomes supersonic, one significant change takes place. The conservation of mass flow rate leads one to expect that contracting the flow channel would increase the flow speed (i.e. making the channel narrower results in faster air flow) and at subsonic speeds this holds true. However, once the flow becomes supersonic, the relationship of flow area and speed is reversed: expanding the channel actually increases the speed.

The obvious result is that in order to accelerate a flow to supersonic, one needs a convergent-divergent nozzle, where the converging section accelerates the flow to sonic speeds, and the diverging section continues the acceleration. Such nozzles are called de Laval nozzles and in extreme cases they are able to reach hypersonic speeds (Mach 13 (15,926 km/h; 9,896 mph) at 20 °C).

An aircraft Machmeter or electronic flight information system (EFIS) can display Mach number derived from stagnation pressure (pitot tube) and static pressure.

## Calculation

The Mach number at which an aircraft is flying can be calculated by

$$M = \frac{u}{c}$$

where:

M is the Mach number

$u$ is velocity of the moving aircraft and

$c$ is the speed of sound at the given altitude

Note that the dynamic pressure can be found as:

$$q = \frac{\gamma}{2} p M^2$$

Assuming air to be an ideal gas, the formula to compute Mach number in a subsonic compressible flow is derived from Bernoulli's equation for M < 1:

$$M = \sqrt{\frac{2}{\gamma - 1}\left[\left(\frac{q_c}{p} + 1\right)^{\frac{\gamma-1}{\gamma}} - 1\right]}$$

where:

$q_c$ is impact pressure (dynamic pressure) and

$p$ is static pressure

$\gamma$ is the ratio of specific heat of a gas at a constant pressure to heat at a constant volume (1.4 for air).

The formula to compute Mach number in a supersonic compressible flow is derived from the Rayleigh supersonic pitot equation:

$$\frac{p_t}{p} = \left[\frac{\gamma+1}{2}M^2\right]^{\frac{\gamma}{\gamma-1}} \cdot \left[\frac{\gamma+1}{1-\gamma+2\gamma M^2}\right]^{\frac{1}{\gamma-1}}$$

## Calculating Mach Number from Pitot Tube Pressure

Mach number is a function of temperature and true airspeed. Aircraft flight instruments, however, operate using pressure differential to compute Mach number, not temperature.

Assuming air to be an ideal gas, the formula to compute Mach number in a subsonic compressible flow is found from Bernoulli's equation for M < 1:

$$= \sqrt{5\left[\left(\frac{}{}+1\right)^{-} -1\right]}$$

The formula to compute Mach number in a supersonic compressible flow can be found from the Rayleigh supersonic pitot equation (above) using parameters for air:

$$M \approx 0.88128485\sqrt{\left(\frac{q_c}{p}+1\right)\left(1-\frac{1}{7M^2}\right)^{2.5}}$$

where:

$q_c$ is the dynamic pressure measured behind a normal shock.

As can be seen, M appears on both sides of the equation, and for practical purposes a root-finding algorithm must be used for a numerical solution. It is first determined whether M is indeed greater than 1.0 by calculating M from the subsonic equation. If M is greater than 1.0 at that point, then the value of M from the subsonic equation is used as the initial condition for fixed point iteration of the supersonic equation, which usually converges in just a few iterations. Alternatively, Newton's method can also be used.

# Supersonic Speed

The term supersonic is used to define a speed that exceeds the speed of sound—a speed that is referred to as Mach 1. However, supersonic airflow is decidedly different from subsonic flow, leading to some debate about the definition of "supersonic." Thus, another definition states that an aircraft (or object) is supersonic only if the airflow over the entire aircraft (or object) is supersonic, which occurs around Mach 1.2 on typical designs. The range Mach 0.75 to 1.2 is considered transonic, when only some parts of the air around the aircraft (or object) reach supersonic speeds. Speeds greater than 5 times the speed of sound are often referred to as hypersonic. Chuck Yeager was the first man to achieve supersonic flight.

## Value of Mach 1

At a typical temperature, such as 21 °C (70 °F), the value of Mach 1 is approximately 344 meters per second (m/s) (1,129 ft/s, 770 mph or 1,238 km/h). Nearly everything about the way an aircraft flies changes dramatically as an aircraft accelerates to supersonic speeds.

## Sound Waves in Gases, Liquids, and Solids

Sounds are traveling vibrations (pressure waves) in an elastic medium. In gases, sound travels longitudinally at different speeds, mostly depending on the molecular mass and temperature of the gas; (pressure has little effect). Because air temperature and composition vary significantly with altitude, Mach numbers for aircraft can change without variation of airspeed. In water at room temperature, supersonic can be considered as any speed greater than 1,440 m/s (4,724 ft/s). In solids, sound waves can be longitudinal or transverse and have even higher velocities.

Supersonic fracture is crack motion faster than the speed of sound in a brittle material.

## Breaking the Sound Barrier

In aerodynamics, the "sound barrier" usually refers to the point at which an aircraft moves from transonic to supersonic speed. The term came into use during World War II, when a number of aircraft started to encounter the effects of compressibility, a grab-bag of unrelated aerodynamic effects, when approaching supersonic speed. The term fell out of use in the 1950s, when aircraft started to routinely "break" the sound barrier.

## Supersonic Objects

The tip of a bullwhip is thought to be the first man-made object to break the sound barrier, resulting in the telltale "crack" (actually a small sonic boom). The wave motion traveling through the bullwhip is what makes it capable of achieving supersonic speeds.

Most modern fighter aircraft are supersonic aircraft, but there have been supersonic passenger aircraft, namely Concorde and the Tupolev Tu-144. Both these passenger aircraft and some modern fighters are also capable of supercruise, a condition of sustained supersonic flight without the use of an afterburner. Due to its ability to supercruise for several hours and the relatively high frequency of flight over several decades, Concorde spent more time flying supersonically than all other aircraft combined by a considerable margin. Since Concorde's final retirement flight on November 26, 2003, there are no supersonic passenger aircraft left in service. Some large bombers, such as the Tupolev Tu-160 and Rockwell B-1 Lancer are also supersonic-capable.

Most modern firearm bullets are supersonic, with rifle projectiles often travelling at speeds approaching and in some cases well exceeding Mach 3.

Most spacecraft, most notably the Space Shuttle are supersonic at least during portions of their reentry, though the effects on the spacecraft are reduced by low air densities. During ascent, launch vehicles generally avoid going supersonic below 30 km (~98,400 feet) to reduce air drag.

Note that the speed of sound decreases somewhat with altitude, due to lower temperatures found there (typically up to 25 km). At even higher altitudes the temperature starts increasing, with the corresponding increase in the speed of sound.

When an inflated balloon is burst, the torn pieces of latex contract at supersonic speed, which contributes to the sharp and loud popping noise.

## Supersonic Flight

Supersonic aerodynamics is simpler than subsonic aerodynamics because the airsheets at different points along the plane often can't affect each other. Supersonic jets and rocket vehicles require several times greater thrust to push through the extra aerodynamic drag experienced within the transonic region (around Mach 0.85–1.2). At these speeds aerospace engineers can gently guide air around the fuselage of the aircraft without producing new shock waves, but any change in cross area farther down the vehicle leads to shock waves along the body. Designers use the Supersonic area rule and the Whitcomb area rule to minimize sudden changes in size.

However, in practical applications, a supersonic aircraft must operate stably in both subsonic and supersonic profiles, hence aerodynamic design is more complex.

One problem with sustained supersonic flight is the generation of heat in flight. At high speeds aerodynamic heating can occur, so an aircraft must be designed to operate and function under very high temperatures. Duralumin, the traditional aircraft material, starts to lose strength and go into plastic deformation at relatively low temperatures, and is unsuitable for continuous use at speeds above Mach 2.2 to 2.4. Materials such as titanium and stainless steel allow operations at much higher temperatures. For example, the Lockheed SR-71 Blackbird jet could fly continuously at Mach 3.1 which could lead to temperatures on some parts of the aircraft getting above 315 °C (600 °F).

Another area of concern for sustained high-speed flight is engine operation. Jet engines create thrust by increasing the temperature of the air they ingest, and as the aircraft speeds up, friction and compression heat this air before it reaches the engines. The maximum allowable temperature of the exhaust is determined by the materials in the turbine at the rear of the engine, so as the aircraft speeds up, the difference in intake and exhaust temperature that the engine can create decreases, and the thrust along with it. Air cooling the turbine area to allow operations at higher temperatures was a key solution, one that continued to improve through the 1950s and on to this day.

Intake design was also a major issue. Normal jet engines can only ingest subsonic air, so for supersonic operation the air must be slowed down. Ramps or cones in the intake are used to create shock waves that slow the airflow before it reaches the engine. Doing so removes energy from the airflow, causing drag. The key to reducing this drag is to use multiple small oblique shock waves, but this was difficult because the angle they make inside the intake changes with Mach number. In order to efficiently operate across a range of speeds, the shock waves have to be "tuned."

An aircraft able to operate for extended periods at supersonic speeds has a potential range advantage over a similar design operating subsonically. Most of the drag an aircraft sees while speeding up to supersonic speeds occurs just below the speed of sound, due to an aerodynamic effect known as wave drag. An aircraft that can accelerate past this speed sees a significant drag decrease, and can fly supersonically with improved fuel economy. However, due to the way lift is generated supersonically, the lift-to-drag ratio of the aircraft as a whole drops, leading to lower range, offsetting or overturning this advantage.

The key to having low supersonic drag is to properly shape the overall aircraft to be long and thin, and close to a "perfect" shape, the von Karman ogive or Sears-Haack body. This has led to almost every supersonic cruising aircraft looking very similar to every other, with a very long and slender fuselage and large delta wings, cf. SR-71, Concorde, etc. Although not ideal for passenger aircraft, this shaping is quite adaptable for bomber use.

## Hypersonic Speed

If travelling faster than sound (otherwise known as supersonic travel) wasn't speedy enough, we're now pushing aircrafts to go even faster: hypersonic speed.

The jump from supersonic to hypersonic travel isn't as clear cut as breaking the sound barrier – there's no sonic boom and instant shift in the aerodynamics – but is loosely defined as "Mach 5", or five times faster than the speed of sound, which would be 6,200km per hour at sea level.

Travel becomes hypersonic when temperatures get so hot that the air molecules become unstable and start losing electrons. At low temperatures, the air is made up of oxygen and nitrogen electrons, but at hypersonic speed, the air begins to change chemically and becomes an electrically-charged field.

Hypersonic speeds have been achieved before, most notably by the U.S. Air Force and NASA when test pilot William J. Knight set the record for fastest crewed flight in 1967 when he piloted the North American X-15 to Mach 6.72 (4,520 mph) at an altitude of 102,100 feet. In 2004, the uncrewed NASA X-43A set a speed record at Mach 9.6, or 7,310 mph at an altitude of about 110,000 feet.

### Changes in the Motion of Objects at Hypersonic Speeds

To put it simply: everything we know about travel gets thrown out the window at hypersonic speeds.

"All of our understanding in aerodynamics falls over," says Dr Bruce. "Every equation that we use to describe sub sonic and supersonic flow and flight – so how we understand pressure force and temperature and the lift and drag of objects – doesn't work when the air starts changing at hypersonic speeds."

Theoretical understanding has dramatically improved in the last 50 years and we now have equations to describe aerodynamics at different Mach levels, but hypersonic travellers are now faced with an engineering challenge.

## Characteristics of Flow

While the definition of hypersonic flow can be quite vague and is generally debatable (especially due to the absence of discontinuity between supersonic and hypersonic flows), a hypersonic flow may be characterized by certain physical phenomena that can no longer be analytically discounted as in supersonic flow. The peculiarity in hypersonic flows are as follows:

1. Shock layer
2. Aerodynamic heating
3. Entropy layer
4. Real gas effects
5. Low density effects
6. Independence of aerodynamic coefficients with Mach number

## Small Shock stand-off Distance

As a body's Mach number increases, the density behind a bow shock generated by the body also increases, which corresponds to a decrease in volume behind the shock due to conservation of mass. Consequently, the distance between the bow shock and the body decreases at higher Mach numbers.

## Entropy Layer

As Mach numbers increase, the entropy change across the shock also increases, which results in a strong entropy gradient and highly vortical flow that mixes with the boundary layer.

## Viscous Interaction

A portion of the large kinetic energy associated with flow at high Mach numbers transforms into internal energy in the fluid due to viscous effects. The increase in internal energy is realized as an increase in temperature. Since the pressure gradient normal to the flow within a boundary layer is approximately zero for low to moderate hypersonic Mach numbers, the increase of temperature through the boundary layer coincides with a decrease in density. This causes the bottom of the boundary layer to expand, so that the boundary layer over the body grows thicker and can often merge with the shock wave near the body leading edge.

## High-temperature Flow

High temperatures due to a manifestation of viscous dissipation cause non-equilibrium chemical flow properties such as vibrational excitation and dissociation and ionization of molecules resulting in convective and radiative heat-flux.

## Classification of Mach Regimes

Although "subsonic" and "supersonic" usually refer to speeds below and above the local speed of sound respectively, aerodynamicists often use these terms to refer to particular ranges of Mach values. This occurs because a "transonic regime" exists around M=1 where approximations of the Navier–Stokes equations used for subsonic design no longer apply, partly because the flow locally exceeds M=1 even when the freestream Mach number is below this value.

The "supersonic regime" usually refers to the set of Mach numbers for which linearised theory may be used; for example, where the (air) flow is not chemically reacting and where heat transfer between air and vehicle may be reasonably neglected in calculations.

Generally, NASA defines "high" hypersonic as any Mach number from 10 to 25, and re-entry speeds as anything greater than Mach 25. Among the aircraft operating in this regime are the Space Shuttle and (theoretically) various developing spaceplanes.

### Regimes

Hypersonic flow can be approximately separated into a number of regimes. The selection of these regimes is rough, due to the blurring of the boundaries where a particular effect can be found.

### Perfect Gas

In this regime, the gas can be regarded as an ideal gas. Flow in this regime is still Mach number dependent. Simulations start to depend on the use of a constant-temperature wall, rather than the adiabatic wall typically used at lower speeds. The lower border of this region is around Mach 5, where ramjets become inefficient, and the upper border around Mach 10-12.

### Two-temperature Ideal Gas

This is a subset of the perfect gas regime, where the gas can be considered chemically perfect, but the rotational and vibrational temperatures of the gas must be considered separately, leading to two temperature models. See particularly the modeling of supersonic nozzles, where vibrational freezing becomes important.

### Dissociated Gas

In this regime, diatomic or polyatomic gases (the gases found in most atmospheres) begin to dissociate as they come into contact with the bow shock generated by the body. Surface catalysis plays a role in the calculation of surface heating, meaning that the type of surface material also has an effect on the flow. The lower border of this regime is where any component of a gas mixture first begins to dissociate in the stagnation point of a flow (which for nitrogen is around 2000 K). At the upper border of this regime, the effects of ionization start to have an effect on the flow.

### Ionized Gas

In this regime the ionized electron population of the stagnated flow becomes significant, and the electrons must be modeled separately. Often the electron temperature is handled separately from the temperature of the remaining gas components. This region occurs for freestream flow velocities around 10–12 km/s. Gases in this region are modeled as non-radiating plasmas.

### Radiation-dominated Regime

Above around 12 km/s, the heat transfer to a vehicle changes from being conductively dominated to radiatively dominated. The modeling of gases in this regime is split into two classes:

1. Optically thin: where the gas does not re-absorb radiation emitted from other parts of the gas,
2. Optically thick: where the radiation must be considered a separate source of energy.

The modeling of optically thick gases is extremely difficult, since, due to the calculation of the radiation at each point, the computation load theoretically expands exponentially as the number of points considered increases.

## Wind Tunnel

Suppose you've just designed a gigantic new passenger airplane and now you want to test it out for real. You could spend millions of dollars building it out of shiny titanium metal and race it down the runway to see if it actually flies—but what if you got your calculations wrong? What if your airplane takes off for twenty seconds, then suddenly drops like a stone and lands on a city packed with 5 million people? It's not the best way for testing something so dangerous. That's why airplane designers try things out on the ground first using scale models in wind tunnels.

Wind tunnels are tube-shaped facilities that allow engineers to move air over a vehicle as if it were flying. They help researchers to learn more about how an aircraft will fly. NASA uses wind tunnels to test scale models of aircraft and spacecraft. Some wind tunnels are big enough to hold full-size versions of vehicles. By moving air around an object, the wind tunnel simulates the conditions of the object in flight.

Designing planes that will fly quickly, efficiently, and economically is all about making air flow smoothly over their wings and past their tube-like bodies. This is called the science of aerodynamics. Once a plane's up in the air, there's no easy way to see how air is moving past it (though an experienced test pilot will have a good idea what might be causing problems). If there's a major design defect, the plane won't get into the air at all. That's why every modern spacecraft and airplane is tested on the ground first in a wind tunnel: a pipe-like building through which air blasts at very high speed.

The basic idea of a wind tunnel is simple: if you can't move the plane through the air, why not move the air past the plane instead? From a scientific point of view, it's exactly the same. If a plane drags (causes air resistance) when it soars through the sky, air will drag in exactly the same way when you fire it past a stationary model of the plane on the ground.

There's nothing to stop you building a super-giant wind tunnel and testing a life-sized model of your plane—and, indeed, the American space agency NASA does have wind tunnels like this. But most of the time it's much cheaper to use a small, scale model of the plane in a much smaller wind tunnel.

## Measurement of Aerodynamic Forces

Air velocity and pressures are measured in several ways in wind tunnels.

Air velocity through the test section is determined by Bernoulli's principle. Measurement of the dynamic pressure, the static pressure, and (for compressible flow only) the temperature rise in the airflow. The direction of airflow around a model can be determined by tufts of yarn attached to the aerodynamic surfaces. The direction of airflow approaching a surface can be visualized by mounting threads in the airflow ahead of and aft of the test model. Smoke or bubbles of liquid can be introduced into the airflow upstream of the test model, and their path around the model can be photographed.

Aerodynamic forces on the test model are usually measured with beam balances, connected to the test model with beams, strings, or cables.

The pressure distributions across the test model have historically been measured by drilling many small holes along the airflow path, and using multi-tube manometers to measure the pressure at each hole. Pressure distributions can more conveniently be measured by the use of pressure-sensitive paint, in which higher local pressure is indicated by lowered fluorescence of the paint at that point. Pressure distributions can also be conveniently measured by the use of pressure-sensitive pressure belts, a recent development in which multiple ultra-miniaturized pressure sensor modules are integrated into a flexible strip. The strip is attached to the aerodynamic surface with tape, and it sends signals depicting the pressure distribution along its surface.

Pressure distributions on a test model can also be determined by performing a wake survey, in which either a single pitot tube is used to obtain multiple readings downstream of the test model, or a multiple-tube manometer is mounted downstream and all its readings are taken.

The aerodynamic properties of an object can not all remain the same for a scaled model. However, by observing certain similarity rules, a very satisfactory correspondence between the aerodynamic properties of a scaled model and a full-size object can be achieved. The choice of similarity parameters depends on the purpose of the test, but the most important conditions to satisfy are usually:

- Geometric similarity: all dimensions of the object must be proportionally scaled;
- Mach number: the ratio of the airspeed to the speed of sound should be identical for the scaled model and the actual object (having identical Mach number in a wind tunnel and around the actual object is -not- equal to having identical airspeeds);
- Reynolds number: the ratio of inertial forces to viscous forces should be kept. This parameter is difficult to satisfy with a scaled model and has led to development of pressurized and cryogenic wind tunnels in which the viscosity of the working fluid can be greatly changed to compensate for the reduced scale of the model.

In certain particular test cases, other similarity parameters must be satisfied, such as e.g. Froude number.

## How it Works

Air is blown or sucked through a duct equipped with a viewing port and instrumentation where models or geometrical shapes are mounted for study. Typically the air is moved through the tunnel using a series of fans. For very large wind tunnels several meters in diameter, a single large fan is not practical, and so instead an array of multiple fans are used in parallel to provide sufficient airflow. Due to the sheer volume and speed of air movement required, the fans may be powered by stationary turbofan engines rather than electric motors.

Six-element external balance below the Kirsten Wind Tunnel

The airflow created by the fans that is entering the tunnel is itself highly turbulent due to the fan blade motion (when the fan is blowing air into the test section – when it is sucking air out of the test section downstream, the fan-blade turbulence is not a factor), and so is not directly useful for accurate measurements. The air moving through the tunnel needs to be relatively turbulence-free and laminar. To correct this problem, closely spaced vertical and horizontal air vanes are used to smooth out the turbulent airflow before reaching the subject of the testing.

Due to the effects of viscosity, the cross-section of a wind tunnel is typically circular rather than square, because there will be greater flow constriction in the corners of a square tunnel that can make the flow turbulent. A circular tunnel provides a smoother flow.

The inside facing of the tunnel is typically as smooth as possible, to reduce surface drag and turbulence that could impact the accuracy of the testing. Even smooth walls induce some drag into the airflow, and so the object being tested is usually kept near the center of the tunnel, with an empty buffer zone between the object and the tunnel walls. There are correction factors to relate wind tunnel test results to open-air results.

The lighting is usually embedded into the circular walls of the tunnel and shines in through windows. If the light were mounted on the inside surface of the tunnel in a conventional manner, the light bulb would generate turbulence as the air blows around it. Similarly, observation is usually done through transparent portholes into the tunnel. Rather than simply being flat discs, these lighting and observation windows may be curved to match the cross-section of the tunnel and further reduce turbulence around the window.

Various techniques are used to study the actual airflow around the geometry and compare it with theoretical results, which must also take into account the Reynolds number and Mach number for the regime of operation.

## Pressure Measurements

Pressure across the surfaces of the model can be measured if the model includes pressure taps. This can be useful for pressure-dominated phenomena, but this only accounts for normal forces on the body.

## Force and Moment Measurements

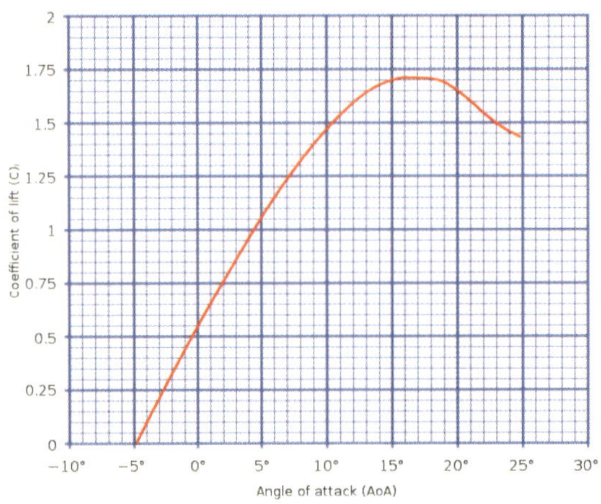


A typical lift coefficient versus angle of attack curve

With the model mounted on a force balance, one can measure lift, drag, lateral forces, yaw, roll, and pitching moments over a range of angle of attack. This allows one to produce common curves such as lift coefficient versus angle of attack.

Note that the force balance itself creates drag and potential turbulence that will affect the model and introduce errors into the measurements. The supporting structures are therefore typically smoothly shaped to minimize turbulence.

## Flow Visualization

Because air is transparent it is difficult to directly observe the air movement itself. Instead, multiple methods of both quantitative and qualitative flow visualization methods have been developed for testing in a wind tunnel.

## Qualitative Methods

- Smoke.
- Tufts are applied to a model and remain attached during testing. Tufts can be used to gauge air flow patterns and flow separation.

- Evaporating suspensions are simply a mixture of some sort or fine powder, talc, or clay mixed into a liquid with a low latent heat of evaporation. When the wind is turned on the liquid quickly evaporates, leaving behind the clay in a pattern characteristic of the air flow.
- Oil: When oil is applied to the model surface it can clearly show the transition from laminar to turbulent flow as well as flow separation.
- Fog (usually from water particles) is created with an ultrasonic piezoelectric nebulizer. The fog is transported inside the wind tunnel (preferably of the closed circuit and closed test section type). An electrically heated grid is inserted before the test section, which evaporates the water particles at its vicinity, thus forming fog sheets. The fog sheets function as streamlines over the test model when illuminated by a light sheet.
- Sublimation: If the air movement in the tunnel is sufficiently non-turbulent, a particle stream released into the airflow will not break up as the air moves along, but stay together as a sharp thin line. Multiple particle streams released from a grid of many nozzles can provide a dynamic three-dimensional shape of the airflow around a body. As with the force balance, these injection pipes and nozzles need to be shaped in a manner that minimizes the introduction of turbulent airflow into the airstream.

High-speed turbulence and vortices can be difficult to see directly, but strobe lights and film cameras or high-speed digital cameras can help to capture events that are a blur to the naked eye.

High-speed cameras are also required when the subject of the test is itself moving at high speed, such as an airplane propeller. The camera can capture stop-motion images of how the blade cuts through the particulate streams and how vortices are generated along the trailing edges of the moving blade.

China clay on a wing in the Kirsten Wind Tunnel showing reverse and span-wise flow

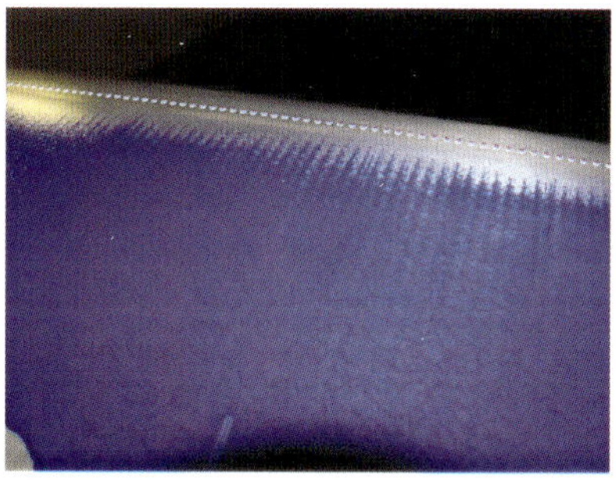

Oil flow visible on a straight wing in the Kirsten Wind Tunnel. Trip dots can be seen near the leading edge.

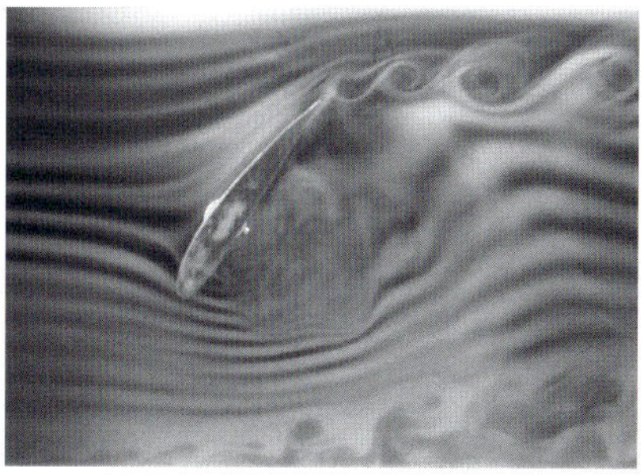

Fog (water particle) wind tunnel visualization of a NACA 4412 airfoil at a low-speed flow (Re=20.000)

## Classification

There are many different kinds of wind tunnels. They are typically classified by the range of speeds that are achieved in the test section, as follows:

- Low-speed wind tunnel
- High speed wind tunnel

- Subsonic and transonic wind tunnel
- Supersonic wind tunnel
- Hypersonic wind tunnel

Wind tunnels are also classified by the orientation of air flow in the test section with respect to gravity. Typically they are oriented horizontally, as happens during level flight. A different class of wind tunnels are oriented vertically so that gravity can be balanced by drag instead of lift, and these have become a popular form of recreation for simulating sky-diving:

- Vertical wind tunnel

Wind tunnels are also classified based on their main use. For those used with land vehicles such as cars and trucks the type of floor aerodynamics is also important. These vary from stationary floors through to full moving floors, with smaller moving floors and some attempt at boundary level control also being important.

## Aeronautical Wind Tunnels

The main subcategories in the aeronautical wind tunnels are:

- High Reynolds Number Tunnels

Reynolds number is one of the governing similarity parameters for the simulation of flow in a wind tunnel. For mach number less than 0.3, it is the primary parameter that governs the flow characteristics. There are three main ways to simulate high Reynolds number, since it is not practical to obtain full scale Reynolds number by use of a full scale vehicle.

- Pressurised tunnels: Here test gases are pressurised to increase the Reynolds number.
- Heavy gas tunnels: Heavier gases like freon and R-134a are used as test gases. The transonic dynamics tunnel at NASA Langley is an example of such a tunnel.
- Cryogenic tunnels: Here test gas is cooled down to increase the Reynolds number. The European transonic wind tunnel uses this technique.
- High-altitude tunnels: These are designed to test the effects of shock waves against various aircraft shapes in near vacuum. In 1952 the University of California constructed the first two high-altitude wind tunnels: one for testing objects at 50 to 70 miles above the earth and the second for tests at 80 to 200 miles above the earth.
- V/STOL Tunnels

V/STOL tunnels require large cross section area, but only small velocities. Since power varies with the cube of velocity, the power required for the operation is also less. An example of a V/STOL tunnel is the NASA Langley 14' x 22' tunnel.

- Spin Tunnels

Aircraft have a tendency to go to spin when they stall. These tunnels are used to study that phenomenon.

## Automotive Tunnels

Automotive wind tunnels fall into two categories:

- External flow tunnels are used to study the external flow through the chassis
- Climatic tunnels are used to evaluate the performance of door systems, braking systems, etc. under various climatic conditions. Most of the leading automobile manufacturers have their own climatic wind tunnels

Wunibald Kamm built the first full-scale wind tunnel for motor vehicles.

For external flow tunnels various systems are used to compensate for the effect of the boundary layer on the road surface, including systems of moving belts under each wheel and the body of the car (5 or 7 belt systems) or one large belt under the entire car, or other methods of boundary layer control such as scoops or perforations to suck it away.

## Aeroacoustic Tunnels

These tunnels are used in the studies of noise generated by flow and its suppression.

## Aquadynamic Flume

Vertical wind tunnel T-105 at Central Aerohydrodynamic Institute, Moscow, built in 1941 for aircraft testing

The aerodynamic principles of the wind tunnel work equally on watercraft, except the water is more viscous and so sets greater forces on the object being tested. A looping flume is typically used for underwater aquadynamic testing. The interaction between two different types of fluids means that pure wind tunnel testing is only partly relevant. However, a similar sort of research is done in a towing tank.

## Low-speed Oversize Liquid Testing

Air is not always the best test medium for studying small-scale aerodynamic principles, due to the speed of the air flow and airfoil movement. A study of fruit fly wings designed to understand how the wings produce lift was performed using a large tank of mineral oil and wings 100 times larger than actual size, in order to slow down the wing beats and make the vortices generated by the insect wings easier to see and understand.

## Fan Testing

Wind tunnel tests are also performed to precisely measure the air movement of fans at a specific pressure. By determining the environmental circumstances during measurement, and by revising the air-tightness afterwards, the standardization of the data is ensured.

There are two possible ways of measurement: a complete fan, or an impeller on a hydraulic installation. Two measuring tubes enable measurements of lower air currents (< 30.000 $m^3/h$) as well as higher air currents (< 60.000 $m^3/h$). The determination of the Q/h curve of the fan is one of the main objectives. To determine this curve (and to define other parameters) air technical, mechanical as well as electro technical data are measured.

Air technical:

- Static pressure difference (Pa)
- Amount of moved air ($m^3/h$)
- Average air speed (m/s)
- Specific efficiency ($W/1000m^3/h$)
- Efficiency

Electro technical:

- Tension (V)
- Current (A)
- Cos φ
- Admitted power (W) fan / impeller
- Rotations per minute (RPM)

The measurement can take place on the fan or in the application in which the fan is used.

## Wind Engineering Testing

In wind engineering, wind tunnel tests are used to measure the velocity around, and forces or pressures upon structures. Very tall buildings, buildings with unusual or complicated shapes (such as a tall building with a parabolic or a hyperbolic shape), cable suspension bridges or cable stayed bridges are analyzed in specialized atmospheric boundary layer wind tunnels. These feature a long upwind section to accurately represent the wind speed and turbulence profile acting on the structure. Wind tunnel tests provide the necessary design pressure measurements in use of the dynamic analysis and control of tall buildings.

## Importance of Wind Tunnels in Spacecraft

NASA also uses wind tunnels to test spacecraft and rockets. These vehicles are made to operate in space. Space has no atmosphere. Spacecraft and rockets have to travel through the atmosphere to get to space. Vehicles that take humans into space also must come back through the atmosphere to Earth.

Wind tunnels are important. NASA uses them to test the Orion spacecraft and the Space Launch System rockets. These rockets are called the SLS. Orion and SLS are new vehicles. They will take astronauts into space. NASA must test the systems in wind tunnels to see if they are safe to fly. And NASA must see what happens when Orion comes back to Earth through the atmosphere.

Wind tunnels can even help engineers design spacecraft to work on other worlds. Mars has a thin atmosphere. It is important to know what the Martian atmosphere will do to vehicles that are landing there. Spacecraft designs and parachutes are tested in wind tunnels set up to be like the Martian atmosphere.

## Types of Wind Tunnels used by NASA

NASA has many different types of wind tunnels. They are located at NASA centers all around the country. The wind tunnels come in a lot of sizes. Some are only a few inches square, and some are large enough to test a full-size airplane. Some wind tunnels test aircraft at very slow speeds. But some wind tunnels are made to test at hypersonic speeds. That is more than 4,000 miles per hour.

## References


- Kirchhoff, G. (1869). "Zur Theorie freier Flussigkeitsstrahlen". Journal für die reine und angewandte Mathematik. 70: 289–298
- G. Falkovich (2011). Fluid Mechanics (A short course for physicists). Cambridge University Press. ISBN 978-1-107-00575-4
- Rayleigh, Lord (1876). "On the Resistance of Fluids". Philosophical Magazine. 2 (13): 430–441. doi:10.1080/14786447608639132
- "U.S Centennial of Flight Commission – Sir George Cayley". Archived from the original on 20 September 2008. Retrieved 2008-09-10
- Spalart, P. R. (2000), Strategies for turbulence modeling and simulations, 21 (3), International Journal of Heat and Fluid Flow, p. 252
- Marchaj, C. A. (2003). Sail performance : techniques to maximise sail power (Rev. ed.). London: Adlard Coles Nautical. pp. 147 figure 127 lift vs drag polar curves. ISBN 978-0-7136-6407-2
- Williamson, C. H. K.; Govardhan, R (2004), "Vortex-induced vibrations", Annual Review of Fluid Mechanics, 36: 413–455, Bibcode:2004AnRFM..36..413W, doi:10.1146/annurev.fluid.36.050802.122128
- Fox, Herbert, and Paul A. Libby. "Some perturbation solutions in laminar boundary layer theory Part 2. The energy equation."Journal of Fluid Mechanics 19.03 (1964): 433-451

- Hernandez-Gomez, J J; Marquina, V; Gomez, R W (25 July 2013). "On the performance of Usain Bolt in the 100 m sprint". Eur. J. Phys. IOP. 34 (5): 1227. Bibcode:2013EJPh...34.1227H. doi:10.1088/0143-0807/34/5/1227. Retrieved 23 April 2016
- Smith, Hubert (1992). The Illustrated Guide to Aerodynamics (2nd ed.). New York: TAB Books. pp. 24–27. ISBN 0-8306-3901-2
- Kármán, Th V. "Über laminare und turbulente Reibung." ZAMM-Journal of Applied Mathematics and Mechanics/Zeitschrift für Angewandte Mathematik und Mechanik 1.4 (1921): 233-252
- Smith, N. F. (1972), "Bernoulli and Newton in Fluid Mechanics", The Physics Teacher, 10 (8): 451, Bibcode:1972PhTea..10..451S, doi:10.1119/1.2352317
- Jr., John D. Anderson,. "Research in Supersonic Flight and the Breaking of the Sound Barrier". history.nasa.gov. Retrieved 14 April 2018
- Roshko, Anatol (1961). "Experiments on the flow past a circular cylinder at very high Reynolds number". Journal of Fluid Mechanics. 10 (3): 345–356. Bibcode:1961JFM....10..345R. doi:10.1017/S0022112061000950
- Young, A.D. (1989). Boundary layers (1st publ. ed.). Washington, DC: American Institute of Aeronautics and Astronautics. ISBN 0930403576
- Wille, R.; Fernholz, H. (1965), "Report on the first European Mechanics Colloquium, on the Coanda effect", J. Fluid Mech., 23 (4): 801–819, Bibcode:1965JFM....23..801W, doi:10.1017/s0022112065001702

3

# Propulsion Systems in Aircrafts

A propulsion system typically consists of a power source like an engine or motor, propellers, wheels and axles, and a propulsive nozzle for generating the force. This chapter discusses the principles of jet propulsion and also elucidates the different innovative designs of aircraft engine and jet engine.

## Jet Propulsion

Jet propulsion revolutionized the science of flight by dramatically increasing possible speeds and altitudes, hence enabling space exploration. The term jet propulsion refers to the action produced by a reactor to the ejection of matter. For example, when the matter in a typical rocket (like gunpowder in fireworks) is ignited, the resulting chemical reaction produces heat and gases, which escape from the rocket and cause it to move forward. The oxygen necessary for combustion is carried (in tanks or in a combined form) in the rocket itself so that the rocket thrust is independent of the atmosphere. Other jet propulsion devices depend on the air inducted into the engine to supply the necessary oxygen. After heat is released by the combustion, the hot gases are accelerated through the engine so that the exit velocity is greater than the airstream velocity at the entrance.

In both self-contained rocket motors and air-breathing jet propulsion power units, the thrust that can be generated is proportional to the mass of material ejected from the unit in a given time, as well as the increase in the mass velocity with respect to the unit. Therefore, the same forward-thrust force can be produced in two ways: by ejecting rearward either a large mass of material at a low velocity during a given time period (as in turbofan engines) or by ejecting a smaller mass of material at higher velocity (as in turbojet and ramjet engines). The two sources of mass are the propellant, or fuel, and the oxidizer, or air.

Fuels contain a large quantity of potential energy, which is rapidly released during combustion. A portion of this heat energy is converted into useful work, moving the vehicle through the atmosphere or into space. Another portion, however, in the form of the jet's kinetic energy, is lost and dissipates into the atmosphere. The very fuel-efficient turbofan engines used in modern commercial aircraft engines attempt to minimize the latter portion. To do so, they impart a moderate increase in the mass velocity to the combustion products for the large mass of air drawn through the engine in a given time. But turbojet and ramjet engines, which meet more demanding supersonic flight requirements, are less fuel-efficient.

Various types of jet engines have been developed to deliver the required thrust and engine performance for a wide range of flight speeds and altitudes. Air-breathing turbojet, turbofan and ramjet engines operate according to similar principles in the sense that they raise the inducted air pressure before combustion and expand the high-energy gases before they leave, in a nozzle or exhaust system. In turbojet engines, the inducted air goes through a compressor to increase its pressure before entering the combustor, then through a turbine before accelerating in the exhaust nozzle. The ramjet engine, though, has no moving parts; it produces a ram pressure rise from decelerating the high-speed inducted air in the inlet diffuser. The ramjet engine can only operate at high supersonic velocities and therefore requires another launching device, such as a rocket or turbojet engine, to accelerate it to the required speed.

Above a certain altitude the atmospheric density diminishes and jet propulsion is only possible for rocket engines that carry their own oxygen. Rocket motors use either solid or liquid fuels. Solid rockets are the oldest types, and their bodies contain the combustion chamber and the solid fuel mixed with oxidizer. When the fuel is ignited the gaseous products of combustion are accelerated through the nozzle to produce thrust. In liquid rockets, the fuel and oxygen are stored in separate tanks and fed at controlled rates to the combustion chamber.

### Physics

Jet propulsion is most effective when the Reynolds number is high - that is, the object being propelled is relatively large and passing through a low-viscosity medium.

In biology, the most efficient jets are pulsed, rather than continuous: at least when the Reynolds number is greater than 6.

### Jet Engine

A jet engine is a reaction engine that discharges a fast moving jet of fluid to generate thrust by jet propulsion and in accordance with Newton's laws of motion. This broad definition of jet engines includes turbojets, turbofans, rockets, ramjets, pulse jets and pump-jets. In general, most jet engines are internal combustion engines but non-combusting forms also exist.

## Aircraft Engine

Aircraft engine means engines (other than those used in military, customs or police services), powered by jet propulsion or turbine or piston technology, and:

- In the case of jet propulsion aircraft engines, have at least 1750 lb of thrust or its equivalent,
- In the case of turbine-powered or piston-powered aircraft engines, have at least 550 rated take-off shaft horsepower or its equivalent,

Together with all modules and other installed, incorporated or attached accessories, parts and equipment,

Aircraft that is further described as aircraft engine must be described by serial number in a registration. The serial number for aircraft engine comprise of,

- The engine's manufacturer's number; that is, the serial number attached to the engine by the manufacturer,

- The manufacturer's name, and
- The manufacturer's generic model description.

## Shaft Engines

### Reciprocating (Piston) Engines

Ranger L-440 air-cooled, six-cylinder, inverted, in-line engine used in Fairchild PT-19

### In-line Engine

In this entry, for clarity, the term "inline engine" refers only to engines with a single row of cylinders, as used in automotive language, but in aviation terms, the phrase "inline engine" also covers V-type and opposed engines, and is not limited to engines with a single row of cylinders. This is typically to differentiate them from radial engines. A straight engine typically has an even number of cylinders, but there are instances of three- and five-cylinder engines. The greatest advantage of an inline engine is that it allows the aircraft to be designed with a low frontal area to minimize drag. If the engine crankshaft is located above the cylinders, it is called an inverted inline engine: this allows the propeller to be mounted high up to increase ground clearance, enabling shorter landing gear. The disadvantages of an inline engine include a poor power-to-weight ratio, because the crankcase and crankshaft are long and thus heavy. An in-line engine may be either air-cooled or liquid-cooled, but liquid-cooling is more common because it is difficult to get enough air-flow to cool the rear cylinders directly. Inline engines were common in early aircraft; one was used in the Wright Flyer, the aircraft that made the first controlled powered flight. However, the inherent disadvantages of the design soon became apparent, and the inline design was abandoned, becoming a rarity in modern aviation.

### V-type Engine

Cylinders in this engine are arranged in two in-line banks, typically tilted 60–90 degrees apart from each other and driving a common crankshaft. The vast majority of V engines are water-cooled. The V design provides a higher power-to-weight ratio than an inline engine, while still providing a small frontal area. Perhaps the most famous example of this design is the legendary Rolls-Royce Merlin engine, a 27-litre (1649 in$^3$) 60° V12 engine used in, among others, the Spitfires that played a major role in the Battle of Britain.

A Rolls-Royce Merlin V-12 Engine

## Horizontally Opposed Engine

A ULPower UL260i horizontally opposed air-cooled aero engine

A horizontally opposed engine, also called a flat or boxer engine, has two banks of cylinders on opposite sides of a centrally located crankcase. The engine is either air-cooled or liquid-cooled, but air-cooled versions predominate. Opposed engines are mounted with the crankshaft horizontal in airplanes, but may be mounted with the crankshaft vertical in helicopters. Due to the cylinder layout, reciprocating forces tend to cancel, resulting in a smooth running engine. Opposed-type engines have high power-to-weight ratios because they have a comparatively small, lightweight crankcase. In addition, the compact cylinder arrangement reduces the engine's frontal area and allows a streamlined installation that minimizes aerodynamic drag. These engines always have an even number of cylinders, since a cylinder on one side of the crankcase "opposes" a cylinder on the other side.

Opposed, air-cooled four- and six-cylinder piston engines are by far the most common engines used in small general aviation aircraft requiring up to 400 horsepower (300 kW) per engine. Aircraft that require more than 400 horsepower (300 kW) per engine tend to be powered by turbine engines.

## H Configuration Engine

An H configuration engine is essentially a pair of horizontally opposed engines placed together, with the two crankshafts geared together.

## Radial Engine

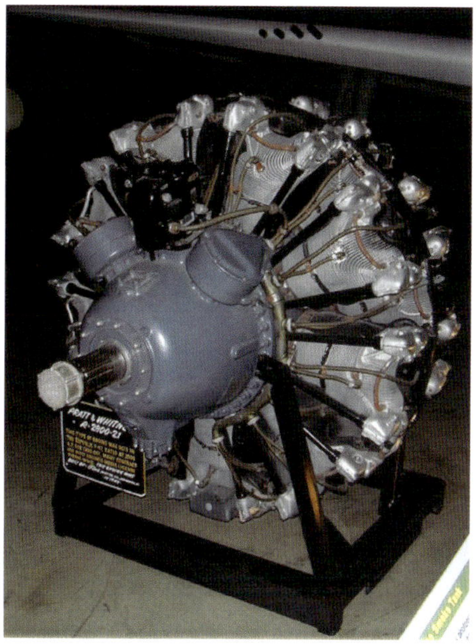

A Pratt & Whitney R-2800 engine

This type of engine has one or more rows of cylinders arranged around a centrally located crankcase. Each row generally has an odd number of cylinders to produce smooth operation. A radial engine has only one crank throw per row and a relatively small crankcase, resulting in a favorable power-to-weight ratio. Because the cylinder arrangement exposes a large amount of the engine's heat-radiating surfaces to the air and tends to cancel reciprocating forces, radials tend to cool evenly and run smoothly. The lower cylinders, which are under the crankcase, may collect oil when the engine has been stopped for an extended period. If this oil is not cleared from the cylinders prior to starting the engine, serious damage due to hydrostatic lock may occur.

Most radial engines have the cylinders arranged evenly around the crankshaft, although some early engines, sometimes called semi-radials or fan configuration engines, had an uneven arrangement. The best known engine of this type is the Anzani engine, which was fitted to the Bleriot XI used for the first flight across the English Channel in 1909. This arrangement had the drawback of needing a heavy counterbalance for the crankshaft, but was used to avoid the spark plugs oiling up.

In military aircraft designs, the large frontal area of the engine acted as an extra layer of armor for the pilot. Also air-cooled engines, without vulnerable radiators, are slightly less prone to battle damage, and on occasion would continue running even with one or more cylinders shot away. However, the large frontal area also resulted in an aircraft with an aerodynamically inefficient increased frontal area.

## Rotary Engine

Rotary engines have the cylinders in a circle around the crankcase, as in a radial engine, but the crankshaft is fixed to the airframe and the propeller is fixed to the engine case, so that the crankcase and cylinders rotate. The advantage of this arrangement is that a satisfactory flow of cooling air is maintained even at low airspeeds, retaining the weight advantage and simplicity of a conventional air-cooled engine without one of their major drawbacks. The first practical rotary engine was the Gnome Omega designed by the Seguin brothers and first flown in 1909. Its relative reliability and good power to weight ratio changed aviation dramatically. Before the first World War most speed records were gained using Gnome-engined aircraft, and in the early years of the war rotary engines were dominant in aircraft types for which speed and agility were paramount. To increase power, engines with two rows of cylinders were built.

Le Rhone 9C rotary aircraft engine

However, the gyroscopic effects of the heavy rotating engine produced handling problems in aircraft and the engines also consumed large amounts of oil since they used total loss lubrication, the oil being mixed with the fuel and ejected with the exhaust gases. Castor oil was used for lubrication, since it is not soluble in petrol, and the resultant fumes were nauseating to the pilots. Engine designers had always been aware of the many limitations of the rotary engine so when the static style engines became more reliable and gave better specific weights and fuel consumption, the days of the rotary engine were numbered.

## Turbine-powered

### Turboprop

Cutaway view of a Garrett TPE-331 turboprop engine showing the gearbox at the front of the engine

While military fighters require very high speeds, many civil airplanes do not. Yet, civil aircraft designers wanted to benefit from the high power and low maintenance that a gas turbine engine offered. Thus was born the idea to mate a turbine engine to a traditional propeller. Because gas turbines optimally spin at high speed, a turboprop features a gearbox to lower the speed of the shaft so that the propeller tips don't reach supersonic speeds. Often the turbines that drive the propeller are separate from the rest of the rotating components so that they can rotate at their own best speed (referred to as a free-turbine engine). A turboprop is very efficient when operated within the realm of cruise speeds it was designed for, which is typically 200 to 400 mph (320 to 640 km/h).

### Turboshaft

An Allison Model 250 turboshaft engine common to many types of helicopters

Turboshaft engines are used primarily for helicopters and auxiliary power units. A turboshaft engine is similar to a turboprop in principle, but in a turboprop the propeller is supported by the engine and the engine is bolted to the airframe: in a turboshaft, the engine does not provide any direct physical support to the helicopter's rotors. The rotor is connected to a transmission which is bolted to the airframe, and the turboshaft engine drives the transmission. The distinction is seen by some as slim, as in some cases aircraft companies make both turboprop and turboshaft engines based on the same design.

## Reaction Engines

Reaction engines generate the thrust to propel an aircraft by ejecting the exhaust gases at high velocity from the engine, the resultant reaction of forces driving the aircraft forwards. The most common reaction propulsion engines flown are turbojets, turbofans and rockets. Other types such as pulsejets, ramjets, scramjets and pulse detonation engines have also flown. In jet engines the oxygen necessary for fuel combustion comes from the air, while rockets carry oxygen in some form as part of the fuel load, permitting their use in space.

### Jets

### Turbojet

A turbojet is a type of gas turbine engine that was originally developed for military fighters during World War II. A turbojet is the simplest of all aircraft gas turbines. It consists of a compressor to draw air in and compress it, a combustion section where fuel is added and ignited, one or more turbines that extract power from the expanding exhaust gases to drive the compressor, and an exhaust nozzle that accelerates the exhaust gases out the back of the engine to create thrust. When turbojets were introduced, the top speed of fighter aircraft equipped with them was at least 100 miles per hour faster than competing piston-driven aircraft. In the years after the war, the drawbacks of the turbojet gradually became apparent. Below about Mach 2, turbojets are very fuel inefficient and create tremendous amounts of noise. Early designs also respond very slowly to power changes, a fact that killed many experienced pilots when they attempted the transition to jets. These drawbacks eventually led to the downfall of the pure turbojet, and only a handful of types are still in production. The last airliner that used turbojets was the Concorde, whose Mach 2 airspeed permitted the engine to be highly efficient.

A General Electric J85-GE-17A turbojet engine. This cutaway clearly shows the 8 stages of axial compressor at the front (left side of the picture), the combustion chambers in the middle, and the two stages of turbines at the rear of the engine.

## Turbofan

A cutaway of a CFM56-3 turbofan engine

A turbofan engine is much the same as a turbojet, but with an enlarged fan at the front that provides thrust in much the same way as a ducted propeller, resulting in improved fuel-efficiency. Though the fan creates thrust like a propeller, the surrounding duct frees it from many of the restrictions that limit propeller performance. This operation is a more efficient way to provide thrust than simply using the jet nozzle alone, and turbofans are more efficient than propellers in the transsonic range of aircraft speeds and can operate in the supersonic realm. A turbofan typically has extra turbine stages to turn the fan. Turbofans were among the first engines to use multiple *spools*—concentric shafts that are free to rotate at their own speed—to let the engine react more quickly to changing power requirements. Turbofans are coarsely split into low-bypass and high-bypass categories. Bypass air flows through the fan, but around the jet core, not mixing with fuel and burning. The ratio of this air to the amount of air flowing through the engine core is the bypass ratio. Low-bypass engines are preferred for military applications such as fighters due to high thrust-to-weight ratio, while high-bypass engines are preferred for civil use for good fuel efficiency and low noise. High-bypass turbofans are usually most efficient when the aircraft is traveling at 500 to 550 miles per hour (800 to 885 km/h), the cruise speed of most large airliners. Low-bypass turbofans can reach supersonic speeds, though normally only when fitted with afterburners.

### Pulse Jets

Pulse jets are mechanically simple devices that—in a repeating cycle—draw air through a no-return valve at the front of the engine into a combustion chamber and ignited it. The combustion forces the exhaust gases out the back of the engine. It produces power as a series of pulses rather than as a steady output, hence the name. The only application of this type of engine was the German unmanned V1 flying bomb of World War II. Though the same engines were also used experimentally for ersatz fighter aircraft, the extremely loud noise generated by the engines caused mechanical damage to the airframe that was sufficient to make the idea unworkable.

### Rocket

An XLR99

A few aircraft have used rocket engines for main thrust or attitude control, notably the Bell X-1 and North American X-15. Rocket engines are not used for most aircraft as the energy and propellant efficiency is very poor, but have been employed for short bursts of speed and takeoff. Where fuel/propellant efficiency is of lesser concern, rocket engines can be useful because they produce very large amounts of thrust and weigh very little.

## Newer Engine Types

### Wankel Engine

Another promising design for aircraft use was the Wankel rotary engine. The Wankel engine is about one half the weight and size of a traditional four-stroke cycle piston engine of equal power output, and much lower in complexity. In an aircraft application, the power-to-weight ratio is very important, making the Wankel engine a good choice. Because the engine is typically constructed with an aluminium housing and a steel rotor, and aluminium expands more than steel when heated, a Wankel engine does not seize when overheated, unlike a piston engine. This is an important safety factor for aeronautical use. Considerable development of these designs started after World War II, but at the time the aircraft industry favored the use of turbine engines. It was believed that turbojet or turboprop engines could power all aircraft, from the largest to smallest designs. The Wankel engine did not find many applications in aircraft, but was used by Mazda in a popular line of sports cars. The French company Citroën had developed Wankel powered RE-2 (fr) helicopter in 1970's.

Powerplant from a Schleicher ASH 26e self-launching motor glider, removed from the glider and mounted on a test stand for maintenance at the Alexander Schleicher GmbH & Co in Poppenhausen, Germany. Counter-clockwise from top left: propeller hub, mast with belt guide, radiator, Wankel engine, muffler shroud.

In modern times the Wankel engine has been used in motor gliders where the compactness, light weight, and smoothness are crucially important.

The now-defunct Staverton-based firm MidWest designed and produced single- and twin-rotor aero engines, the MidWest AE series. These engines were developed from the motor in the Norton Classic motorcycle. The twin-rotor version was fitted into ARV Super2s and the Rutan Quickie. The single-rotor engine was put into a Chevvron motor glider and into the Schleicher ASH motor-gliders. After the demise of MidWest, all rights were sold to Diamond of Austria, who have since developed a MkII version of the engine.

As a cost-effective alternative to certified aircraft engines some Wankel engines, removed from automobiles and converted to aviation use, have been fitted in homebuilt experimental aircraft. Mazda units with outputs ranging from 100 horsepower (75 kW) to 300 horsepower (220 kW) can be a fraction of the cost of traditional engines. Such conversions first took place in the early 1970s; and as of 10 December 2006 the National Transportation Safety Board has only seven reports of incidents involving aircraft with Mazda engines, and none of these is of a failure due to design or manufacturing flaws.

## Diesel Engine

Most aircraft engines use spark ignition, generally using gasoline as a fuel. Starting in the 1930s attempts were made to produce a compression ignition Diesel engine for aviation use. In general, Diesel engines are more reliable and much better suited to running for long periods of time at medium power settings, which is why they are widely used in, for example, trucks and ships. The lightweight alloys of the 1930s were not up to the task of handling the much higher compression ratios of diesel engines, so they generally had poor power-to-weight ratios and were uncommon for that reason, although the Clerget 14F Diesel radial engine (1939) has the same power to weight ratio as a gasoline radial. Improvements in Diesel technology in automobiles (leading to much better power-weight ratios), the Diesel's much better fuel efficiency and the high relative taxation of AVGAS compared to Jet A1 in Europe have all seen a revival of interest in the use of diesels for aircraft. Thielert Aircraft Engines converted Mercedes Diesel automotive engines, certified them for aircraft use, and became an OEM provider to Diamond Aviation for their light twin. Financial problems have plagued Thielert, so Diamond's affiliate — Austro Engine — developed the new AE300 turbodiesel, also based on a Mercedes engine. Competing new Diesel engines may bring fuel efficiency and lead-free emissions to small aircraft, representing the biggest change in light aircraft engines in decades.

## Precooled Jet Engines

For very high supersonic/low hypersonic flight speeds, inserting a cooling system into the air duct of a hydrogen jet engine permits greater fuel injection at high speed and obviates the need for the duct to be made of refractory or actively cooled materials. This greatly improves the thrust/weight ratio of the engine at high speed.

It is thought that this design of engine could permit sufficient performance for antipodal flight at Mach 5, or even permit a single stage to orbit vehicle to be practical. The hybrid air-breathing SABRE rocket engine is a pre-cooled engine under development.

## Electric

A number of electrically powered aircraft, such as the QinetiQ Zephyr, have been designed since the 1960s. Some are used as military drones. In France in late 2007, a conventional light aircraft powered by an 18 kW electric motor using lithium polymer batteries was flown, covering more than 50 kilometers (31 mi), the first electric airplane to receive a certificate of airworthiness.

Limited experiments with solar electric propulsion have been performed, notably the manned Solar Challenger and Solar Impulse and the unmanned NASA Pathfinder aircraft.

Many big companies, as Siemens, are developing high performance electric engines for aircraft use, also, SAE shows new developments in elements as pure Copper core electric motors with a better efficiency.

Small multicopter UAVs are almost always powered by electric motors.

## Piston-turbofan Hybrid

At the April 2018 ILA Berlin Air Show, Munich-based research institute de:Bauhaus Luftfahrt presented a high-efficiency composite cycle engine for 2050, combining a geared turbofan with a piston engine core. The 2.87 m diameter, 16-blade fan gives a 33.7 ultra-high bypass ratio, powered by a geared low-pressure turbine but the high-pressure compressor is driven by a piston-engine with two 10 piston banks without a high-pressure turbine, increasing efficiency with non-stationary isochoric-isobaric combustion for higher peak pressures and temperatures. The 11,200 lb (49.7 kN) engine could power a 50-seat regional jet.

Its cruise TSFC would be 11.5 g/kN/s (0.406 lb/lbf/hr) for an overall engine efficiency of 48.2%, for a burner temperature of 1,700 K (1,430 °C), an overall pressure ratio of 38 and a peak pressure of 30 MPa (300 bar). Although engine weight increases by 30%, aircraft fuel consumption is reduced by 15%. Sponsored by the European Commission under Framework 7 project LEMCOTEC, Bauhaus Luftfahrt, MTU Aero Engines and GKN Aerospace presented the concept in 2015, raising the overall engine pressure ratio to over 100 for a 15.2% fuel burn reduction compared to 2025 engines.

## Engine Position Numbering

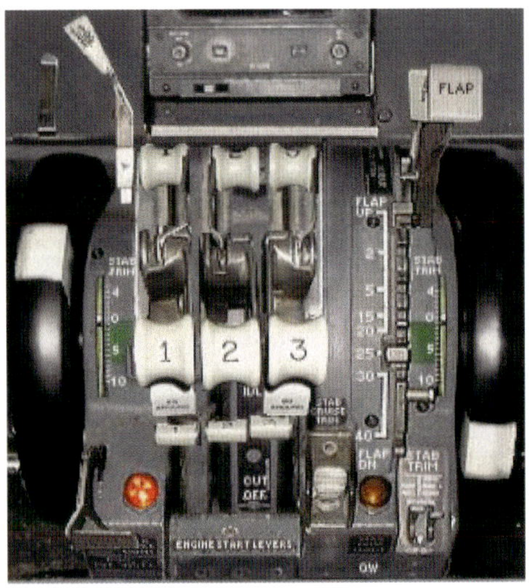


The thrust levers of a three-engine Boeing 727, each one bearing the respective engine number

On multi-engine aircraft, engine positions are numbered from left to right from the point of view of the pilot looking forward, so for example on a four-engine aircraft such as the Boeing 747, engine No. 1 is on the left side, farthest from the fuselage, while engine No. 3 is on the right side nearest to the fuselage.

In the case of the twin-engine English Electric Lightning, which has two fuselage-mounted jet engines one above the other, engine No. 1 is below and to the front of engine No. 2, which is above and behind.

In the Cessna 337 Skymaster, a push-pull twin-engine airplane, engine No. 1 is the one at the front of the fuselage, while engine No. 2 is aft of the cabin.

## Fuel

Aircraft reciprocating (piston) engines are typically designed to run on aviation gasoline. Avgas has a higher octane rating than automotive gasoline to allow higher compression ratios, power output, and efficiency at higher altitudes. Currently the most common Avgas is 100LL. This refers to the octane rating (100 octane) and the lead content (LL = low lead, relative to the historic levels of lead in pre-regulation Avgas).

Refineries blend Avgas with tetraethyllead (TEL) to achieve these high octane ratings, a practice that governments no longer permit for gasoline intended for road vehicles. The shrinking supply of TEL and the possibility of environmental legislation banning its use have made a search for replacement fuels for general aviation aircraft a priority for pilots' organizations.

Turbine engines and aircraft Diesel engines burn various grades of jet fuel. Jet fuel is a relatively less volatile petroleum derivative based on kerosene, but certified to strict aviation standards, with additional additives.

Model aircraft typically use nitro engines (also known as "glow engines" due to the use of a glow plug) powered by glow fuel, a mixture of methanol, nitromethane, and lubricant. Electrically powered model airplanes and helicopters are also commercially available. Small multicopter UAVs are almost always powered by electricity, but larger gasoline-powered designs are under development.

## Propfan

It is described as "a small diameter, highly loaded multiple bladed variable pitch propulsor having swept blades with thin advanced airfoil sections, integrated with a nacelle contoured to retard the airflow through the blades thereby reducing compressibility losses and designed to operate with a turbine engine and using a single stage reduction gear resulting in high performance." The design is intended to offer the speed and performance of a turbofan, with the fuel economy of a turboprop. The propfan concept was first revealed by Carl Rohrbach and Bruce Metzger of the Hamilton Standard Division of United Technologies in 1975 and was patented by Robert Cornell and Carl Rohrbach of Hamilton Standard in 1979. Later work by General Electric on similar propulsors was done under the name unducted fan, which was a modified turbofan engine, with the fan placed outside the engine nacelle on the same axis as the compressor blades.

The European Aviation Safety Agency (EASA) defines it as "*A turbine engine featuring contra-rotating fan stages not enclosed within a casing.*" The engine uses a gas turbine to drive an unshrouded (open) contra-rotating propeller like a turboprop, but the design of the propeller itself is more tightly coupled to the turbine design and the two are certified as a single unit.

United Technologies describe it as "a small diameter, highly loaded multiple bladed variable pitch propulsor having swept blades with thin advanced airfoil sections, integrated with a nacelle contoured to retard the airflow through the blades thereby reducing compressibility losses and designed to operate with a turbine engine and using a single stage reduction gear resulting in high performance."

## Limitations and Solutions

### Propeller Blade tip Speed Limit

Turboprops have an optimum speed below about 450 mph (700 km/h). The reason is that all propellers lose efficiency at high speed, due to an effect known as wave drag that occurs just below supersonic speeds. This powerful form of drag has a sudden onset, and led to the concept of a sound barrier when it was first encountered in the 1940s. In the case of a propeller, this effect can happen any time the propeller is spun fast enough that the blade tips approach the speed of sound, even if the aircraft is motionless on the ground.

The most effective way to counteract this problem (to some degree) is by adding more blades to the propeller, allowing it to deliver more power at a lower rotational speed. This is why many World War II fighter designs started with two or three-blade propellers and by the end of the war were using up to five blades in some cases as the engines were upgraded and new propellers were needed to more efficiently convert that power. The major downside to this approach is that adding blades makes the propeller harder to balance and maintain and the additional blades cause minor performance penalties (due to drag and efficiency issues). But even with these sorts of measures at some point the forward speed of the plane combined with the rotational speed of the propeller will once again result in wave drag problems. For most aircraft this will occur at speeds over about 450 mph (700 km/h).

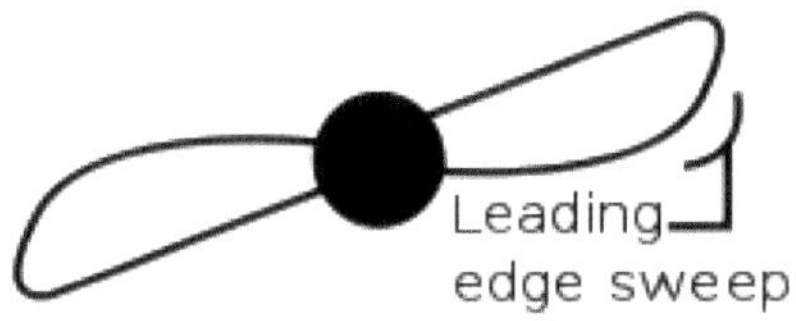


Swept propeller

A method of decreasing wave drag was discovered by German researchers in 1935—sweeping the wing backwards. Today, almost all aircraft designed to fly much above 450 mph (700 km/h) use a swept wing. In the 1970s, Hamilton Standard started researching propellers with similar sweep. Since the inside of the propeller is moving slower than the outside, the blade is progressively more swept toward the outside, leading to a curved shape similar to a scimitar - a practice that was first used as far back as 1909, in the Chauvière make of a two-bladed wood propeller used on the Blériot XI.

## Jet Aircraft fuel Economy

Jet aircraft fly faster than conventional propeller-driven aircraft. However, they use more fuel, so that for the same fuel consumption a propeller installation produces more thrust. As fuel costs become an increasingly important aspect of commercial aviation, engine designers continue to seek ways to improve aero engine efficiency.

The propfan concept was developed to deliver 35% better fuel efficiency than contemporary turbofans. In static and air tests on a modified Douglas DC-9, propfans reached a 30% improvement over the OEM turbofans. This efficiency came at a price, as one of the major problems with the propfan is noise, particularly in an era where aircraft are required to comply with increasingly strict aircraft noise regulations.

## Aircraft with Propfans

- Antonov An-70
- McDonnell Douglas MD-94X

## Antonov An-70

Antonov An-70 is a four-engine medium military freighter aircraft designed and manufactured by Antonov ASTC of Ukraine. The short take-off and landing (STOL) capable aircraft is being built to replace the company's An-12 Cub transport aircraft. It is the first large aircraft to be powered by propfan engines. Two An-70 prototypes have been built to date.

## An-70 Cargo Aircraft Variants, Orders and Deliveries

The An-70 has three variants, including the An-70T, the An-70TK and the An-112KC.

The An-70T is a cargo transport plane principally used for commercial applications. The An-70TK is a twin engine passenger or freight transformable aircraft. The An-112KC is an aerial refuelling model used in the KC-X programme for the USAF.

The Ukrainian Air Force ordered five An-70 aircraft in November 2005. The fuselage of the first serial production aircraft was completed in December 2012. Deliveries are expected to begin in 2013.

The Russian Air Force ordered 40 An-70s in June 2010. Deliveries are expected to begin in 2013.

Five An-70s will be procured by Volga-Dnepr airlines. Antonov signed a contract with the Czech Republic in 2005 to supply three An-70 military transport airplanes.

Russia and Ukraine are planning to procure 165 and 65 An-70 aircraft respectively during the next 20 years.

Russia may purchase An-70 freighters in 2013 based on the success of flight tests to be conducted in 2013. It is currently operating a fleet of An-12 Cub, IL-76MD and An-124 transport aircraft.

## Antonov An-70 Design and Cockpit

The An-70 was designed to take-off from and land on concrete and unpaved airstrips including soft ground and snow. It can manoeuvre day or night in bad weather conditions.

The airframe is constructed with composite materials to reduce the overall weight of the aircraft by 20% to 30%.

The aircraft boasts a large freight compartment for cargo storage. The compartment is 19.1m long, four metres wide and 4.1m high. It can accommodate 35,000kg to 47,000kg of cargo and 300 troops. It can also carry 38,000kg of fuel.

The An-70 features a spacious glass cockpit which can accommodate the three crew members – pilot, co-pilot and flight engineer. The cockpit is fitted with six full colour digital displays and a head up display, which aids the aircraft's landing on short airfields.

Automated navigational and flight control systems are incorporated for quick analysis and servicing.

## Engines and Cargo Systems of the Military Freighter Aircraft

The Antonov An-70 is powered by four Ivchenko Progress D-27 three-shaft propfan engines. Each engine can produce 10,350kW of output power. FADEC compatible, the D-27 engine is fitted with two remote gearboxes, a single stage compressor, a four stage low pressure turbine and a jetpipe.

*Antonov An-70 is a four-engine medium military freighter aircraft.*

The engine is 4.19m long, 1.25m wide and 1.37m high. Its dry weight is 1,650kg. The lifespan of the engine is 18,000 hours.

The incorporation of SV-27 counter rotating propeller in the power plant increases cruise speed as well as reduces the fuel consumption by 20% to 30% as compared to modern turbojets.

The aircraft features four overhead rail electric motor hoists to lift 12,000kg of cargo. Two on-board electric winches can lift 30,000kg of freight. A built-in aerial delivery system allows autonomous loading or unloading of cargoes and can be used for air drops.

### Antonov An-70 Performance

The An-70 can climb at a rate of 24.9m/s. The maximum and cruise speeds of the aircraft are 780km/h and 750km/h respectively.

The stall speed is 113km/h. The range of the aircraft is 6,600km, while the ferry range is 8,000km. The service ceiling is 12,000m.

### McDonnell Douglas MD-94X

This is the McDonnell-Douglas MD-94X, a proposed MD-90 variant with the PW/Allison 578-DX Propfan engines. Never entered commercial service, as the project was canceled due to lack of interest by airlines into a propfan aircraft. This aircraft was modified from Skycycle's 717, with permission, and contains the 717 Cockpit panel and modified livery. landing at 120kts.

## Jet Engine

Turn your eyes to the sky and it's likely you'll see more than a few vapor trails—the wispy white lines that jet planes scribble on the great blue canvas stretched above our heads. At the dawn of the 20th century, the very idea of powered flight seemed, to many, like a ludicrous dream. How things have changed! At any given moment, there are something like 7,000 flights zipping through the sky over the United States alone; we're so used to the idea of flight that we barely even notice all the planes screaming above us, hauling hundreds of people at a time to their homes and holidays. Most modern planes are powered by jet engines. What exactly are these magic machines and what makes them different from the engines used in cars or trucks?

A jet engine is a machine that converts energy-rich, liquid fuel into a powerful pushing force called thrust. The thrust from one or more engines pushes a plane forward, forcing air past its scientifically shaped wings to create an upward force called lift that powers it into the sky.

One way to understand modern jet engines is to compare them with the piston engines used in early airplanes, which are very similar to the ones still used in cars. A piston engine (also called a reciprocating engine, because the pistons move back and forth or "reciprocate") makes its power in strong steel "cooking pots" called cylinders. Fuel is squirted into the cylinders with air from the atmosphere. The piston in each cylinder compresses the mixture, raising its temperature so it either ignites spontaneously (in a diesel engine) or with help from a sparking plug (in a gas engine). The burning fuel and air explodes and expands, pushing the piston back out and driving the crankshaft that powers the car's wheels (or the plane's propeller), before the whole four-step cycle (intake, compression, combustion, exhaust) repeats itself. The trouble with this is that the piston is driven only during one of the four steps—so it's making power only a fraction of the time. The amount of power a piston engine makes is directly related to how big the cylinder is and how far the piston moves; unless you use hefty cylinders and pistons (or many of them), you're limited to producing relatively modest amounts of power. If your piston engine is powering a plane, that limits how fast it can fly, how much lift it can make, how big it can be, and how much it can carry.

## Uses

A JT9D turbofan jet engine installed on a Boeing 747 aircraft.

Jet engines power jet aircraft, cruise missiles and unmanned aerial vehicles. In the form of rocket engines they power fireworks, model rocketry, spaceflight, and military missiles.

Jet engines have propelled high speed cars, particularly drag racers, with the all-time record held by a rocket car. A turbofan powered car, ThrustSSC, currently holds the land speed record.

Jet engine designs are frequently modified for non-aircraft applications, as industrial gas turbines or marine powerplants. These are used in electrical power generation, for powering water, natural gas, or oil pumps, and providing propulsion for ships and locomotives. Industrial gas turbines can create up to 50,000 shaft horsepower. Many of these engines are derived from older military turbojets such as the Pratt & Whitney J57 and J75 models. There is also a derivative of the P&W JT8D low-bypass turbofan that creates up to 35,000 HP.

Jet engines are also sometimes developed into, or share certain components such as engine cores, with turboshaft and turbopropengines, which are forms of gas turbine engines that are typically used to power helicopters and some propeller-driven aircraft.

## Types of Jet Engine

There are a large number of different types of jet engines, all of which achieve forward thrust from the principle of *jet propulsion*.

### Airbreathing

Commonly aircraft are propelled by airbreathing jet engines. Most airbreathing jet engines that are in use are turbofan jet engines, which give good efficiency at speeds just below the speed of sound.

### Turbine Powered

Gas turbines are rotary engines that extract energy from a flow of combustion gas. They have an upstream compressor coupled to a downstream turbine with a combustion chamber in-between. In aircraft engines, those three core components are often called the "gas generator." There are many different variations of gas turbines, but they all use a gas generator system of some type.

### Turbojet

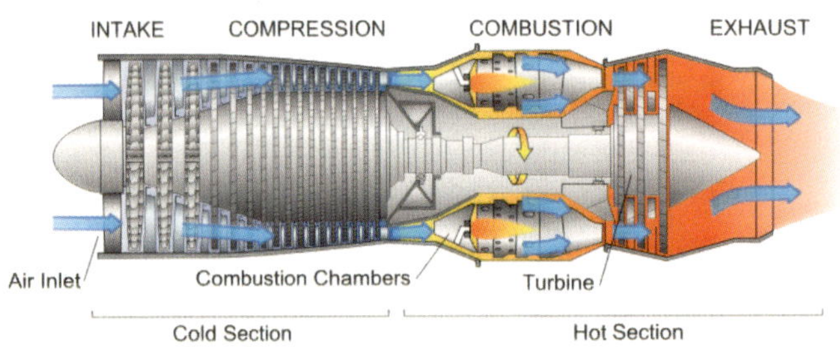


Turbojet engine

A turbojet engine is a gas turbine engine that works by compressing air with an inlet and a compressor (axial, centrifugal, or both), mixing fuel with the compressed air, burning the mixture in the combustor, and then passing the hot, high pressure air through a turbine and a nozzle. The compressor is powered by the turbine, which extracts energy from the expanding gas passing through it. The engine converts internal energy in the fuel to kinetic energy in the exhaust, producing thrust. All the air ingested by the inlet is passed through the compressor, combustor, and turbine, unlike the turbofan engine.

### Turbofan

Turbofans differ from turbojets in that they have an additional fan at the front of the engine, which accelerates air in a duct bypassing the core gas turbine engine. Compared to a turbojet of identical thrust, a turbofan has a much larger air mass flow rate. Turbofans are the dominant engine type for medium and long-range airliners.

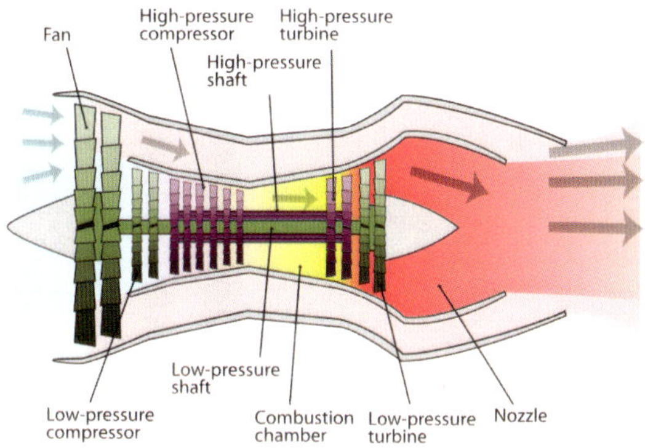

Schematic diagram illustrating the operation of a low-bypass turbofan engine.

The comparatively large frontal fan has several effects. The main effect is that the output of the engine as a whole has a much higher mass per second, and therefore generates much more thrust, despite not having ignited much of its airflow. Also, because the additional air has not been ignited, no extra fuel is needed to provide this thrust. The slower average velocity of the mixed exhaust air (low specific thrust) is also less wasteful of energy for subsonic flight, and allows the engine to be more efficient and much quieter, while the fan also allows greater thrust to be available at slow speeds. Together, the thrust produced by the fan and core are much more fuel efficient, and provides a much higher output, than could be produced by the core alone.

Turbofans are usually more efficient than turbojets at subsonic speeds, but their large frontal area also generates more drag.Therefore, in supersonic flight, and in military and other aircraft where absolute performance, weight, and drag have a higher priority than fuel efficiency, engines tend to either have smaller fans (or multiple smaller fans) or use other engine designs entirely. Typically, turbofans in civilian aircraft usually have a pronounced large front area to accommodate a very large fan, as their design involves a much larger mass of air bypassing the core so they can benefit from these effects, while in military aircraft, where noise and efficiency are less important compared to performance and drag, a smaller amount of air typically bypasses the core. Turbofans designed for subsonic civilian aircraft also usually have a just a single front fan, because their additional thrust is generated by a large additional mass of air which is only moderately compressed, rather than a smaller amount of air which is greatly compressed.

Because of these distinctions, turbofan engine designs are often categorized as low-bypass or high-bypass, depending upon the amount of air which bypasses the core of the engine. Low-bypass turbofans have a bypass ratio of around 2:1 or less, meaning that for each kilogram of air that passes through the core of the engine, two kilograms or less of air bypass the core. Low-bypass turbofans often use a mixed exhaust nozzle meaning that the bypassed flow and the core flow exit from the same nozzle. High-bypass turbofans often have ratios from 4:1 up to 8:1, with the Rolls-Royce Trent XWB approaching 10:1.

## Turboprop and Turboshaft

Turboprop engines are jet engine derivatives, still gas turbines, that extract work from the hot-exhaust jet to turn a rotating shaft, which is then used to produce thrust by some other means. While not strictly jet engines in that they rely on an auxiliary mechanism to produce thrust, turboprops are very similar to other turbine-based jet engines, and are often described as such.

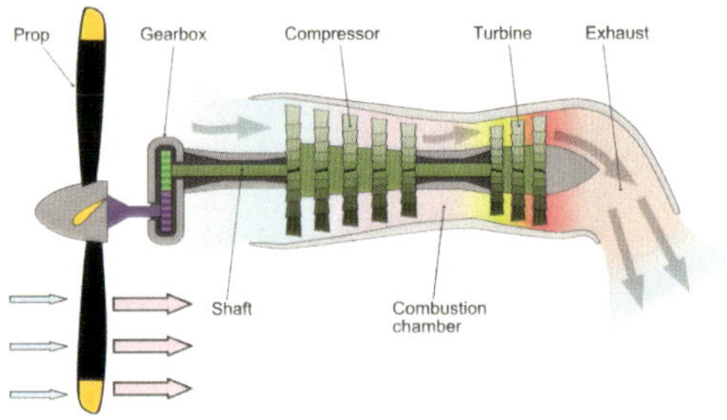


Turboprop engine

In turboprop engines, a portion of the engine's thrust is produced by spinning a propeller, rather than relying solely on high-speed jet exhaust. As their jet thrust is augmented by a propeller, turboprops are occasionally referred to as a type of hybrid jet engine. They are quite similar to turbofans in many respects, except that they use a traditional propeller to provide the majority of thrust, rather than a ducted fan. Both fans and propellers are powered the same way, although most turboprops use gear-reduction between the turbine and the propeller (geared turbofans also feature gear reduction). While many turboprops generate the majority of their thrust with the propeller, the hot-jet exhaust is an important design point, and maximum thrust is obtained by matching thrust contributions of the propeller to the hot jet. Turboprops generally have better performance than turbojets or turbofans at low speeds where propeller efficiency is high, but become increasingly noisy and inefficient at high speeds.

Turboshaft engines are very similar to turboprops, differing in that nearly all energy in the exhaust is extracted to spin the rotating shaft, which is used to power machinery rather than a propeller, they therefore generate little to no jet thrust and are often used to power helicopters.

## Propfan

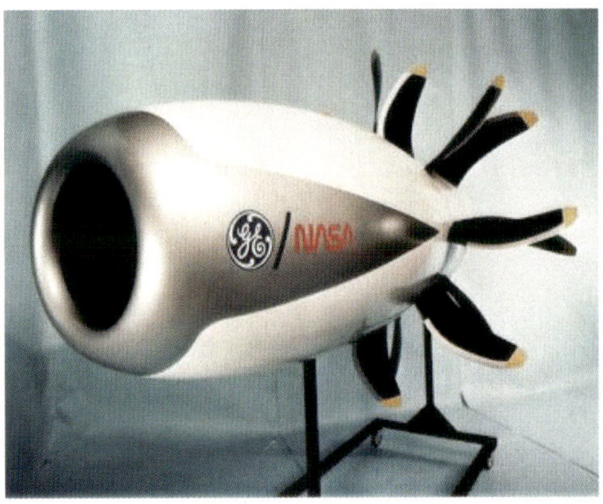


A propfan engine

A propfan engine (also called "unducted fan", "open rotor", or "ultra-high bypass") is a jet engine that uses its gas generator to power an exposed fan, similar to turboprop engines. Like turboprop engines, propfans generate most of their thrust from the propeller and not the exhaust jet. The primary difference between turboprop and propfan design is that the propeller blades on a propfan are highly swept to allow them to operate at speeds around Mach 0.8, which is competitive with modern commercial turbofans. These engines have the fuel efficiency advantages of turboprops with the performance capability of commercial turbofans. While significant research and testing (including flight testing) has been conducted on propfans, none have entered production.

## Ram Compression

Ram compression jet engines are airbreathing engines similar to gas turbine engines and they both follow the Brayton cycle. Gas turbine and ram powered engines differ, however, in how they compress the incoming airflow. Whereas gas turbine engines use axial or centrifugal compressors to compress incoming air, ram engines rely only on air compressed through the inlet or diffuser. A ram engine thus requires a substantial initial forward airspeed before it can function. Ram powered engines are considered the most simple type of air breathing jet engine because they can contain no moving parts.

### Ramjet

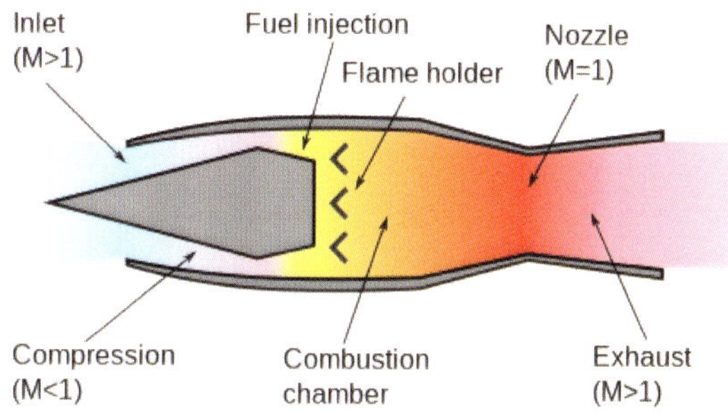


A schematic of a ramjet engine, where "M" is the Mach number of the airflow.

Ramjets are the most basic type of ram powered jet engines. They consist of three sections; an inlet to compress incoming air, a combustor to inject and combust fuel, and a nozzle to expel the hot gases and produce thrust. Ramjets require a relatively high speed to efficiently compress the incoming air, so ramjets cannot operate at a standstill and they are most efficient at supersonic speeds. A key trait of ramjet engines is that combustion is done at subsonic speeds. The supersonic incoming air is dramatically slowed through the inlet, where it is then combusted at the much slower, subsonic, speeds. The faster the incoming air is, however, the less efficient it becomes to slow it to subsonic speeds. Therefore, ramjet engines are limited to approximately Mach 5.

### Scramjet

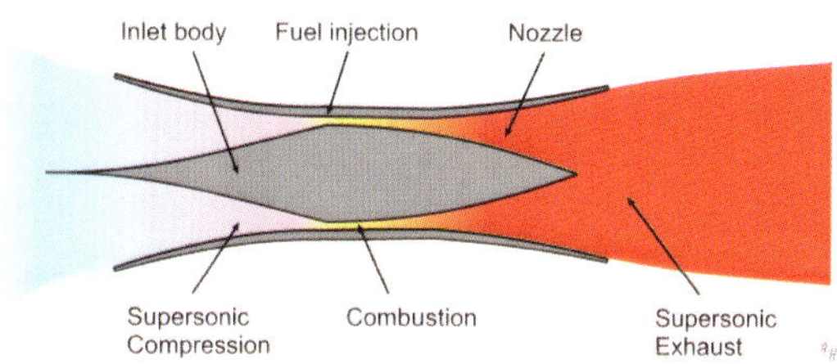


Scramjet engine operation

Scramjets are mechanically very similar to ramjets. Like a ramjet, they consist of an inlet, a combustor, and a nozzle. The primary difference between ramjets and scramjets is that scramjets do not slow the oncoming airflow to subsonic speeds for combustion, they use supersonic combustion instead. The name "scramjet" comes from "Supersonic Combusting Ramjet." Since scramjets use supersonic combustion they can operate at speeds above Mach 6 where traditional ramjets are too inefficient. Another difference between ramjets and scramjets comes from how each type of engine compresses the oncoming airflow: while the inlet provides most of the compression for ramjets, the high speeds at which scramjets operate allow them to take advantage of the compression generated by shock waves, primarily oblique shocks.

Very few scramjet engines have ever been built and flown. In May 2010 the Boeing X-51 set the endurance record for the longest scramjet burn at over 200 seconds.

## Non-continuous Combustion

| Type | Description | Advantages | Disadvantages |
|---|---|---|---|
| **Motorjet** | Works like a turbojet but instead of a turbine driving the compressor a piston engine drives it. | Higher exhaust velocity than a propeller, offering better thrust at high speed | Heavy, inefficient and underpowered. Example: Caproni Campini N.1. |
| **Pulsejet** | Air is compressed and combusted intermittently instead of continuously. Some designs use valves. | Very simple design, used for the V-1 flying bomb and more recently on model aircraft | Noisy, inefficient (low compression ratio), works poorly on a large scale, valves on valved designs wear out quickly |
| **Pulse detonation engine** | Similar to a pulsejet, but combustion occurs as a detonation instead of a deflagration, may or may not need valves | Maximum theoretical engine efficiency | Extremely noisy, parts subject to extreme mechanical fatigue, hard to start detonation, not practical for current use |

## General Physical Principles

All jet engines are reaction engines that generate thrust by emitting a jet of fluid rearwards at relatively high speed. The forces on the inside of the engine needed to create this jet give a strong thrust on the engine which pushes the craft forwards.

Jet engines make their jet from propellant from tankage that is attached to the engine (as in a 'rocket') as well as in duct engines(those commonly used on aircraft) by ingesting an external fluid (very typically air) and expelling it at higher speed.

## Propelling Nozzle

The propelling nozzle is the key component of all jet engines as it creates the exhaust jet. Propelling nozzles turn internal and pressure energy into high velocity kinetic energy. The total pressure and temperature don't change through the nozzle but their static values drop as the gas speeds up.

The velocity of the air entering the nozzle is low, about Mach 0.4, a prerequisite for minimizing pressure losses in the duct leading to the nozzle. The temperature entering the nozzle may be as low as sea level ambient for a fan nozzle in the cold air at cruise altitudes. It may be as high as the 1000K exhaust gas temperature for a supersonic afterburning engine or 2200K with afterburner lit. The pressure entering the nozzle may vary from 1.5 times the pressure outside the nozzle, for a single stage fan, to 30 times for the fastest manned aircraft at mach 3+.

The velocity of the gas leaving a convergent nozzle gives a sonic (Mach 1) exhaust which may be suitable for subsonic flight speeds but for higher speeds a supersonic a con-di nozzle is needed.

The nozzle thrust is highest if the static pressure of the gas reaches the ambient value as it leaves the nozzle. This only happens if the nozzle exit area is the correct value for the nozzle pressure ratio (npr). Since the npr changes with engine thrust setting and flight speed this is seldom the case. Also at supersonic speeds the divergent area is less than required to give complete internal expansion to ambient pressure as a trade-off with external body drag. Whitford gives the F-16 as an example. Other underexpanded examples were the XB-70 and SR-71.

The nozzle size, together with the area of the turbine nozzles, determines the operating pressure of the compressor.

## Thrust

### Origin of Engine Thrust

The familiar explanation for jet thrust is a "black box" description which only looks at what goes into the engine, air and fuel, and what comes out, exhaust gas and an unbalanced force. This force, called thrust, is the sum of the momentum difference between entry and exit and any unbalanced pressure force between entry and exit, as explained in "Thrust calculation". As an example, an early turbojet, the Bristol Olympus Mk. 101, had a momentum thrust of 9300 lb. and a pressure thrust of 1800 lb. giving a total of 11,100 lb. Looking inside the "black box" shows that the thrust results from all the unbalanced momentum and pressure forces created within the engine itself. These forces, some forwards and some rearwards, are across all the internal parts, both stationary and rotating, such as ducts, compressors, etc., which are in the primary gas flow which flows through the engine from front to rear. The algebraic sum of all these forces is delivered to the airframe for propulsion. "Flight" gives examples of these internal forces for two early jet engines, the Rolls-Royce Avon Ra.14 and the de Havilland Goblin.

### Transferring Thrust to the Aircraft

The engine thrust acts along the engine centreline. The aircraft "holds" the engine on the outer casing of the engine at some distance from the engine centreline (at the engine mounts). This arrangement causes the engine casing to bend (known as backbone bending) and the round rotor casings to distort (ovalization). Distortion of the engine structure has to be controlled with suitable mount locations to maintain acceptable rotor and seal clearances and prevent rubbing. A well-publicized example of excessive structural deformation occurred with the original Pratt & Whitney JT9D engine installation in the Boeing 747 aircraft. The engine mounting arrangement had to be revised with the addition of an extra thrust frame to reduce the casing deflections to an acceptable amount.

### Rotor Thrust

The rotor thrust on a thrust bearing is not related to the engine thrust. It may even change direction at some RPM. The bearing load is determined by bearing life considerations. Although the aerodynamic loads on the compressor and turbine blades contribute to the rotor thrust they are small compared to cavity loads inside the rotor which result from the secondary air system pressures and sealing diameters on discs, etc. To keep the load within the bearing specification seal diameters are chosen accordingly as, many years ago, on the backface of the impeller in the de Havilland Ghost engine. Sometimes an extra disc known as a balance piston has to be added inside the rotor. An early turbojet example with a balance piston was the Rolls-Royce Avon.

## Thrust Calculation

The net thrust ($\mathbf{F_N}$) of an engine is given by::p16

$$F_N = (\dot{m}_{air} + \dot{m}_{fuel})v_e - \dot{m}_{air}v$$

| where: | |
|---|---|
| $\dot{m}_{air}$ | = the mass rate of air flow through the engine |
| $\dot{m}_{fuel}$ | = the mass rate of propellant flow entering the engine |
| $v_e$ | = the effective exhaust velocity of the jet (the speed of the exhaust plume relative to the aircraft) |
| $v$ | = the velocity of the air intake = the true airspeed of the aircraft |
| $(\dot{m}_{air} + \dot{m}_{fuel})v_e$ | = the nozzle gross thrust ($F_G$) |
| $\dot{m}_{air}v$ | = the ram drag of the intake air |

Most types of jet engine have an air intake, which provides the bulk of the fluid exiting the exhaust. Conventional rocket engines, however, do not have an intake, so $\dot{m}_{air}$ is zero. Therefore, rocket engines do not have ram drag and the gross thrust of the rocket engine nozzle is the net thrust of the engine. Consequently, the thrust characteristics of a rocket motor are different from that of an air breathing jet engine, and thrust is independent of velocity.

If the velocity of the jet from a jet engine is equal to sonic velocity, the jet engine's nozzle is said to be choked. If the nozzle is choked, the pressure at the nozzle exit plane is greater than atmospheric pressure, and extra terms must be added to the above equation to account for the pressure thrust.

The rate of flow of fuel entering the engine is often very small compared with the rate of flow of air. When the contribution of fuel to the nozzle gross thrust can be ignored, the net thrust is:

$$F_N = \dot{m}_{air}(v_e - v)$$

The velocity of the jet ($v_e$) must exceed the true airspeed of the aircraft ($v$) if there is to be a net forward thrust on the aircraft. The velocity ($v_e$) can be calculated thermodynamically based on adiabatic expansion.

## Thrust Augmentation

Thrust augmentation has taken many forms, most commonly to supplement inadequate take-off thrust. Some early jet aircraft needed rocket assistance to take off from high altitude airfields or when the day temperature was high. A more recent aircraft, the Tupolev Tu-22 supersonic bomber, was fitted with four SPRD-63 boosters for take-off. Possibly the most extreme requirement needing rocket assistance, and which was short-lived, was zero-length launching. Almost as extreme, but very common, is catapult assistance from aircraft carriers. Rocket assistance has also been used during flight. The SEPR 841 booster engine was used on the Dassault Miragefor high altitude interception.

Early aft-fan arrangements which added bypass airflow to a turbojet were known as thrust augmentors. The aft-fan fitted to the General Electric CJ805-3 turbojet augmented the take-off thrust from 11,650lb to 16,100lb.

Water, or other coolant, injection into the compressor or combustion chamber and fuel injection into the jetpipe (afterburning/reheat) became standard ways to increase thrust, known as 'wet' thrust to differentiate with the no-augmentation 'dry' thrust.

Coolant injection (pre-compressor cooling) has been used, together with afterburning, to increase thrust at supersonic speeds. The 'Skyburner' McDonnell Douglas F-4 Phantom II set a world speed record using water injection in front of the engine.

At high Mach numbers afterburners supply progressively more of the engine thrust as the thrust from the turbomachine drops off towards zero at which speed the engine pressure ratio (epr) has fallen to 1.0 and all the engine thrust comes from the afterburner. The afterburner also has to make up for the pressure loss across the turbomachine which is a drag item at higher speeds where the epr will be less than 1.0.

Thrust augmentation of existing afterburning engine installations for special short-duration tasks has been the subject of studies for launching small payloads into low earth orbits using aircraft such as McDonnell Douglas F-4 Phantom II, McDonnell Douglas F-15 Eagle, Dassault Rafale and Mikoyan MiG-31, and also for carrying experimental packages to high altitudes using a Lockheed SR-71. In the first case an increase in the existing maximum speed capability is required for orbital launches. In the second case an increase in thrust within the existing speed capability is required. Compressor inlet cooling is used in the first case. A compressor mapshows that the airflow reduces with increasing compressor inlet temperature although the compressor is still running at maximum RPM (but reduced aerodynamic speed). Compressor inlet cooling increases the aerodynamic speed and flow and thrust. In the second case a small increase in the maximum mechanical speed and turbine temperature were allowed, together with nitrous oxide injection into the afterburner and simultaneous increase in afterburner fuel flow.

## Energy Efficiency Relating to Aircraft Jet Engines

This overview highlights where energy losses occur in complete jet aircraft powerplants or engine installations.

A jet engine at rest, as on a test stand, sucks in fuel and tries to thrust itself forward. How well it does this is judged by how much fuel it uses and what force is required to restrain it. This is a measure of its efficiency. If something deteriorates inside the engine (known as performance deterioration) it will be less efficient and this will show when the fuel produces less thrust. If a change is made to an internal part which allows the air/combustion gases to flow more smoothly the engine will be more efficient and use less fuel. A standard definition is used to assess how different things change engine efficiency and also to allow comparisons to be made between different engines. This definition is called specific fuel consumption, or how much fuel is needed to produce one unit of thrust. For example, it will be known for a particular engine design that if some bumps in a bypass duct are smoothed out the air will flow more smoothly giving a pressure loss reduction of x% and y% less fuel will be needed to get the take-off thrust, for example. This understanding comes under the engineering discipline Jet engine performance. How efficiency is affected by forward speed and by supplying energy to aircraft systems is mentioned later.

The efficiency of the engine is controlled primarily by the operating conditions inside the engine which are the pressure produced by the compressor and the temperature of the combustion gases at the first set of rotating turbine blades. The pressure is the highest air pressure in the engine. The turbine rotor temperature is not the highest in the engine but is the highest at which energy transfer takes place ( higher temperatures occur in the combustor). The above pressure and temperature are shown on a Thermodynamic cycle diagram.

The efficiency is further modified by how smoothly the air and the combustion gases flow through the engine, how well the flow is aligned (known as incidence angle) with the moving and stationary passages in the compressors and turbines. Non-optimum angles, as well as non-optimum passage and blade shapes can cause thickening and separation of Boundary layers and formation of Shock waves. It is important to slow the flow (lower speed means less pressure losses or Pressure drop) when it travels through ducts connecting the different parts. How well the individual components contribute to turning fuel into thrust is quantified by measures like efficiencies for the compressors, turbines and combustor and pressure losses for the ducts. These are shown as lines on a Thermodynamic cycle diagram.

The engine efficiency, or thermal efficiency, known as $\eta_{th}$. is dependent on the Thermodynamic cycle parameters, maximum pressure and temperature, and on component efficiencies, $\eta_{compressor}$, $\eta_{combustion}$ and $\eta_{turbine}$ and duct pressure losses.

The engine needs compressed air for itself just to run successfully. This air comes from its own compressor and is called secondary air. It does not contribute to making thrust so makes the engine less efficient. It is used to preserve the mechanical integrity of the engine, to stop parts overheating and to prevent oil escaping from bearings for example. Only some of this air taken from the compressors returns to the turbine flow to contribute to thrust production. Any reduction in the amount needed improves the engine efficiency. Again, it will be known for a particular engine design that a reduced requirement for cooling flow of x% will reduce the specific fuel consumption by y%. In other words, less fuel will be required to give take-off thrust, for example. The engine is more efficient.

All of the above considerations are basic to the engine running on its own and, at the same time, doing nothing useful, i.e. it is not moving an aircraft or supplying energy for the aircraft's electrical, hydraulic and air systems. In the aircraft the engine gives away some of its thrust-producing potential, or fuel, to power these systems. These requirements, which cause installation losses, reduce its efficiency. It is using some fuel that does not contribute to the engine's thrust.

Finally, when the aircraft is flying the propelling jet itself contains wasted kinetic energy after it has left the engine. This is quantified by the term propulsive, or Froude, efficiency $\eta_p$ and may be reduced by redesigning the engine to give it bypass flow and a lower speed for the propelling jet, for example as a turboprop or turbofan engine. At the same time forward speed increases the $\eta_{th}$ by increasing the Overall pressure ratio.

The overall efficiency of the engine at flight speed is defined as $\eta_o = \eta_p \eta_{th}$.

The $\eta_o$ at flight speed depends on how well the intake compresses the air before it is handed over to the engine compressors. The intake compression ratio, which can be as high as 32:1 at Mach 3, adds to that of the engine compressor to give the Overall pressure ratio and $\eta_{th}$ for the Thermodynamic cycle. How well it does this is defined by its pressure recovery or measure of the losses in the intake. Mach 3 manned flight has provided an interesting illustration of how these losses can increase dramatically in an instant. The North American XB-70 Valkyrie and Lockheed SR-71 Blackbird at Mach 3 each had pressure recoveries of about 0.8, due to relatively low losses during the compression process, i.e. through systems of multiple shocks. During an 'unstart' the efficient shock system would be replaced by a very inefficient single shock beyond the inlet and an intake pressure recovery of about 0.3 and a correspondingly low pressure ratio.

The propelling nozzle at speeds above about Mach 2 usually has extra internal thrust losses because the exit area is not big enough as a trade-off with external afterbody drag.

Although a bypass engine improves propulsive efficiency it incurs losses of its own inside the engine itself. Machinery has to be added to transfer energy from the gas generator to a bypass airflow. The low loss from the propelling nozzle of a turbojet is added to with extra losses due to inefficiencies in the added turbine and fan. These may be included in a transmission, or transfer, efficiency $\eta_T$. However, these losses are more than made up by the improvement in propulsive efficiency. There are also extra pressure losses in the bypass duct and an extra propelling nozzle.

With the advent of turbofans with their loss-making machinery what goes on inside the engine has been separated by Bennett, for example, between gas generator and transfer machinery giving $\eta_o = \eta_p \eta_{th} \eta_T$.

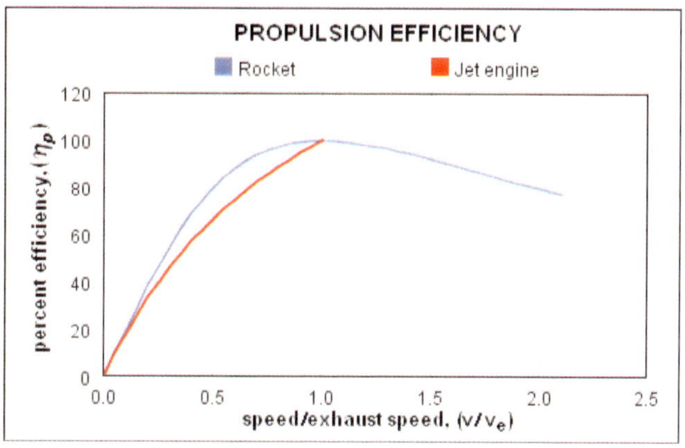


Dependence of propulsion efficiency (η) upon the vehicle speed/exhaust velocity ratio ($v/v_e$) for air-breathing jet and rocket engines.

The energy efficiency $(\eta_o)$ of jet engines installed in vehicles has two main components:

- *Propulsive efficiency* ($\eta_p$): how much of the energy of the jet ends up in the vehicle body rather than being carried away as kinetic energy of the jet.
- *Cycle efficiency* ($\eta_{th}$): how efficiently the engine can accelerate the jet.

Even though overall energy efficiency $\eta_o$ is:

$$\eta_o = \eta_p \eta_{th}$$

for all jet engines the *propulsive efficiency* is highest as the exhaust jet velocity gets closer to the vehicle speed as this gives the smallest residual kinetic energy. For an airbreathing engine an exhaust velocity equal to the vehicle velocity, or a $\eta_p$ equal to one, gives zero thrust with no net momentum change. The formula for air-breathing engines moving at speed with an exhaust velocity , and neglecting fuel flow, is:

$$\eta_p = \frac{2}{1 + \frac{v_e}{v}}$$

And for a rocket:

$$\eta_p = \frac{2(\frac{v}{v_e})}{1 + (\frac{v}{v_e})^2}$$

In addition to propulsive efficiency, another factor is *cycle efficiency*; a jet engine is a form of heat engine. Heat engine efficiency is determined by the ratio of temperatures reached in the engine to that exhausted at the nozzle. This has improved constantly over time as new materials have been introduced to allow higher maximum cycle temperatures. For example, composite materials, combining metals with ceramics, have been developed for HP turbine blades, which run at the maximum cycle temperature. The efficiency is also limited by the overall pressure ratio that can be achieved. Cycle efficiency is highest in rocket engines (~60+%), as they can achieve extremely high combustion temperatures. Cycle efficiency in turbojet and similar is nearer to 30%, due to much lower peak cycle temperatures.

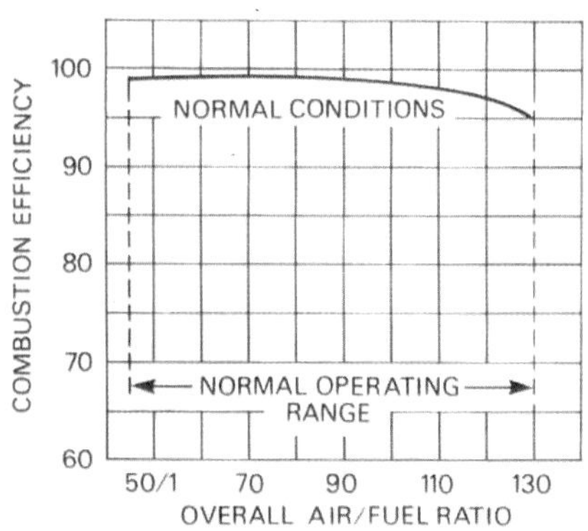


Typical combustion efficiency of an aircraft gas turbine over the operational range

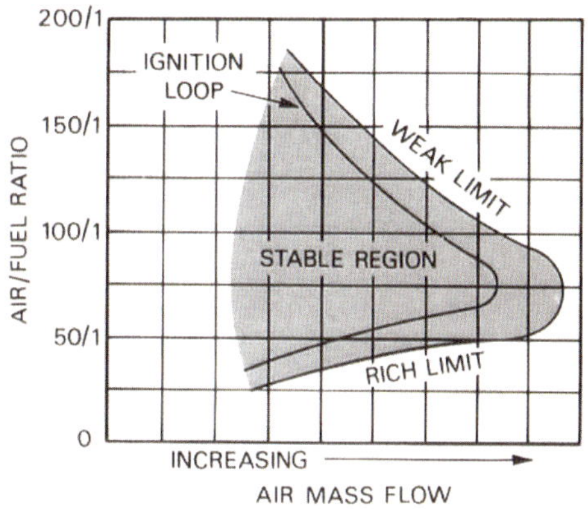


Typical combustion stability limits of an aircraft gas turbine

The combustion efficiency of most aircraft gas turbine engines at sea level takeoff conditions is almost 100%. It decreases nonlinearly to 98% at altitude cruise conditions. Air-fuel ratio ranges from 50:1 to 130:1. For any type of combustion chamber there is a *rich* and *weak limit* to the air-fuel ratio, beyond which the flame is extinguished. The range of air-fuel ratio between the rich and weak limits is reduced with an increase of air velocity. If the increasing air mass flow reduces the fuel ratio below certain value, flame extinction occurs.

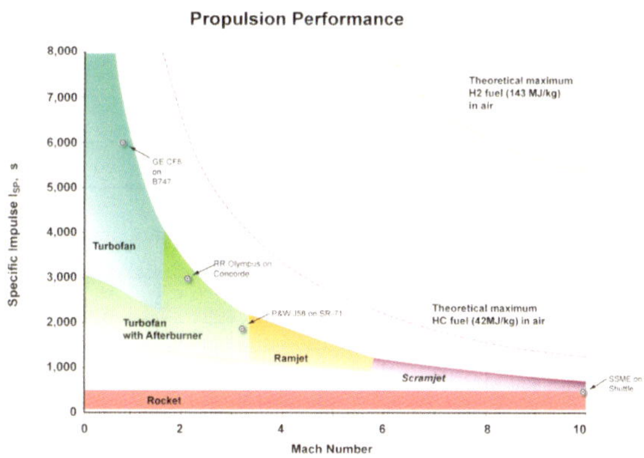


Specific impulse as a function of speed for different jet types with kerosene fuel (hydrogen $I_{sp}$ would be about twice as high). Although efficiency plummets with speed, greater distances are covered. Efficiency per unit distance (per km or mile) is roughly independent of speed for jet engines as a group; however, airframes become inefficient at supersonic speeds

## Consumption of Fuel or Propellant

A closely related (but different) concept to energy efficiency is the rate of consumption of propellant mass. Propellant consumption in jet engines is measured by Specific Fuel Consumption, Specific impulse or Effective exhaust velocity. They all measure the same thing. Specific impulse and effective exhaust velocity are strictly proportional, whereas specific fuel consumption is inversely proportional to the others.

For airbreathing engines such as turbojets, energy efficiency and propellant (fuel) efficiency are much the same thing, since the propellant is a fuel and the source of energy. In rocketry, the propellant is also the exhaust, and this means that a high energy propellant gives better propellant efficiency but can in some cases actually give *lower* energy efficiency.

It can be seen in the table (just below) that the subsonic turbofans such as General Electric's CF6 turbofan use a lot less fuel to generate thrust for a second than did the Concorde's Rolls-Royce/ Snecma Olympus 593 turbojet. However, since energy is force times distance and the distance per second was greater for Concorde, the actual power generated by the engine for the same amount of fuel was higher for Concorde at Mach 2 than the CF6. Thus, the Concorde's engines were more efficient in terms of energy per mile.

**Specific fuel consumption (SFC), specific impulse, and effective exhaust velocity numbers for various rocket and jet engines.**

| Engine type | Scenario | Spec. fuel cons. | | Specific impulse (s) | Effective exhaust velocity (m/s) |
|---|---|---|---|---|---|
| | | (lb/lb-f·h) | (g/kN·s) | | |
| NK-33 rocket engine | Vacuum | 10.9 | 308 | 331 | 3250 |
| SSME rocket engine | Space shuttle vacuum | 7.95 | 225 | 453 | 4440 |
| Ramjet | Mach 1 | 4.5 | 130 | 800 | 7800 |
| J-58 turbojet | SR-71 at Mach 3.2 (Wet) | 1.9 | 54 | 1900 | 19000 |
| Eurojet EJ200 | Reheat | 1.66–1.73 | 47–49 | 2080–2170 | 20400–21300 |
| Rolls-Royce/Snecma Olympus 593 turbojet | Concorde Mach 2 cruise (Dry) | 1.195 | 33.8 | 3010 | 29500 |
| Eurojet EJ200 | Dry | 0.74–0.81 | 21–23 | 4400–4900 | 44000–48000 |
| CF6-80C2B1F turbofan | Boeing 747-400 cruise | 0.605 | 17.1 | 5950 | 58400 |
| General Electric CF6 turbofan | Sea level | 0.307 | 8.7 | 11700 | 115000 |

## Thrust-to-weight Ratio

The thrust-to-weight ratio of jet engines with similar configurations varies with scale, but is mostly a function of engine construction technology. For a given engine, the lighter the engine, the better the thrust-to-weight is, the less fuel is used to compensate for drag due to the lift needed to carry the engine weight, or to accelerate the mass of the engine.

As can be seen in the following table, rocket engines generally achieve much higher thrust-to-weight ratios than duct engines such as turbojet and turbofan engines. This is primarily because rockets almost universally use dense liquid or solid reaction mass which gives a much smaller volume and hence the pressurization system that supplies the nozzle is much smaller and lighter for the same performance. Duct engines have to deal with air which is two to three orders of magnitude less dense and this gives pressures over much larger areas, which in turn results in more engineering materials being needed to hold the engine together and for the air compressor.

| Jet or rocket engine | Mass | | Thrust (vacuum) | | Thrust-to-weight ratio |
|---|---|---|---|---|---|
| | (kg) | (lb) | (kN) | (lbf) | |
| RD-0410 nuclear rocket engine | 2,000 | 4,400 | 35.2 | 7,900 | 1.8 |
| J58 jet engine (SR-71 Blackbird) | 2,722 | 6,001 | 150 | 34,000 | 5.2 |
| Rolls-Royce/Snecma Olympus 593 turbojet with reheat (Concorde) | 3,175 | 7,000 | 169.2 | 38,000 | 5.4 |
| Pratt & Whitney F119 | 1,800 | 3,900 | 91 | 20,500 | 7.95 |
| RD-0750 rocket engine, three-propellant mode | 4,621 | 10,188 | 1,413 | 318,000 | 31.2 |
| RD-0146 rocket engine | 260 | 570 | 98 | 22,000 | 38.4 |
| Space Shuttle Main Engine | 3,177 | 7,004 | 2,278 | 512,000 | 73.1 |
| RD-180 rocket engine | 5,393 | 11,890 | 4,152 | 933,000 | 78.5 |

| | | | | | |
|---|---|---|---|---|---|
| RD-170 rocket engine | 9,750 | 21,500 | 7,887 | 1,773,000 | 82.5 |
| F-1 (Saturn V first stage) | 8,391 | 18,499 | 7,740.5 | 1,740,100 | 94.1 |
| NK-33 rocket engine | 1,222 | 2,694 | 1,638 | 368,000 | 136.7 |
| Merlin 1D rocket engine, full-thrust version | 467 | 1,030 | 825 | 185,000 | 180.1 |

## Comparison of Types

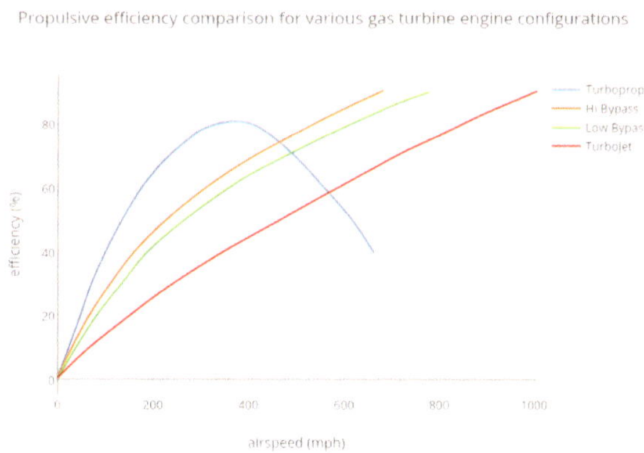


Propulsive efficiency comparison for various gas turbine engine configurations

Propeller engines handle larger air mass flows, and give them smaller acceleration, than jet engines. Since the increase in air speed is small, at high flight speeds the thrust available to propeller-driven aeroplanes is small. However, at low speeds, these engines benefit from relatively high propulsive efficiency.

On the other hand, turbojets accelerate a much smaller mass flow of intake air and burned fuel, but they then reject it at very high speed. When a de Laval nozzle is used to accelerate a hot engine exhaust, the outlet velocity may be locally supersonic. Turbojets are particularly suitable for aircraft travelling at very high speeds.

Turbofans have a mixed exhaust consisting of the bypass air and the hot combustion product gas from the core engine. The amount of air that bypasses the core engine compared to the amount flowing into the engine determines what is called a turbofan's bypass ratio (BPR).

While a turbojet engine uses all of the engine's output to produce thrust in the form of a hot high-velocity exhaust gas jet, a turbofan's cool low-velocity bypass air yields between 30% and 70% of the total thrust produced by a turbofan system.

The net thrust ($F_N$) generated by a turbofan can also be expanded as:

$$F_N = \dot{m}_e v_{he} - \dot{m}_o v_o + BPR(\dot{m}_c v_f)$$

where:

| | |
|---|---|
| $\dot{m}_e$ | = the mass rate of hot combustion exhaust flow from the core engine |
| $\dot{m}_o$ | = the mass rate of total air flow entering the turbofan = $\dot{m}_c + \dot{m}_f$ |
| $\dot{m}_c$ | = the mass rate of intake air that flows to the core engine |
| $\dot{m}_f$ | = the mass rate of intake air that bypasses the core engine |
| $v_f$ | = the velocity of the air flow bypassed around the core engine |

| | |
|---|---|
| $v_{he}$ | = the velocity of the hot exhaust gas from the core engine |
| $v_o$ | = the velocity of the total air intake = the true airspeed of the aircraft |
| *BPR* | = Bypass Ratio |

Rocket engines have extremely high exhaust velocity and thus are best suited for high speeds (hypersonic) and great altitudes. At any given throttle, the thrust and efficiency of a rocket motor improves slightly with increasing altitude (because the back-pressure falls thus increasing net thrust at the nozzle exit plane), whereas with a turbojet (or turbofan) the falling density of the air entering the intake (and the hot gases leaving the nozzle) causes the net thrust to decrease with increasing altitude. Rocket engines are more efficient than even scramjets above roughly Mach 15.

## Altitude and Speed

With the exception of scramjets, jet engines, deprived of their inlet systems can only accept air at around half the speed of sound. The inlet system's job for transonic and supersonic aircraft is to slow the air and perform some of the compression.

The limit on maximum altitude for engines is set by flammability- at very high altitudes the air becomes too thin to burn, or after compression, too hot. For turbojet engines altitudes of about 40 km appear to be possible, whereas for ramjet engines 55 km may be achievable. Scramjets may theoretically manage 75 km. Rocket engines of course have no upper limit.

At more modest altitudes, flying faster compresses the air at the front of the engine, and this greatly heats the air. The upper limit is usually thought to be about Mach 5-8, as above about Mach 5.5, the atmospheric nitrogen tends to react due to the high temperatures at the inlet and this consumes significant energy. The exception to this is scramjets which may be able to achieve about Mach 15 or more, as they avoid slowing the air, and rockets again have no particular speed limit.

## Noise

The noise emitted by a jet engine has many sources. These include, in the case of gas turbine engines, the fan, compressor, combustor, turbine and propelling jet/s.

The propelling jet produces jet noise which is caused by the violent mixing action of the high speed jet with the surrounding air. In the subsonic case the noise is produced by eddies and in the supersonic case by Mach waves. The sound power radiated from a jet varies with the jet velocity raised to the eighth power for velocities up to 2,000 ft/sec and varies with the velocity cubed above 2,000 ft/sec. Thus, the lower speed exhaust jets emitted from engines such as high bypass turbofans are the quietest, whereas the fastest jets, such as rockets, turbojets, and ramjets, are the loudest. For commercial jet aircraft the jet noise has reduced from the turbojet through bypass engines to turbofans as a result of a progressive reduction in propelling jet velocities. For example, the JT8D, a bypass engine, has a jet velocity of 1450 ft/sec whereas the JT9D, a turbofan, has jet velocities of 885 ft/sec (cold) and 1190 ft/sec (hot).

The advent of the turbofan replaced the very distinctive jet noise with another sound known as "buzz saw" noise. The origin is the shockwaves originating at the supersonic fan blades at takeoff thrust.

## Airbreathing Jet Engine

A jet engine sucks air in the front and then pushes it out the back. This is called an air-breathing engine. A jet engine uses a lot of gas.

The air going through the engine can turn a propeller or a fan. This kind of engine uses less gas. Some air-breathing engines use a propeller, some do not.

All the air-breathing airplanes can fly very fast. Very fast hot air comes out the back of the airplane. This is called an air-breathing engine.

All practical air breathing engines are internal combustion engines that directly heat the air by burning fuel, with the resultant hot gases used for propulsion via a propulsive nozzle. A continuous stream of air flows through the air-breathing engine. The air is compressed, mixed with fuel, ignited and expelled as the exhaust gas. Thrust produced by a typical air-breathing engine is about eight times greater than its weight. The thrust results from the expulsion of the working gases from the exhaust nozzle. To expel the gases from the nozzle at high velocity, the air entering the combustion chamber of the engine is compressed.

## Types of Airbreathing Jet Engines

Airbreathing jet engines are nearly always internal combustion engines that obtain propulsion from the combustion of fuel inside the engine. Oxygen present in the atmosphere is used to oxidise a fuel source, typically a hydrocarbon-based jet fuel. The burning mixture expands greatly in volume, driving heated air through a propelling nozzle.

Gas turbine powered engines:

- turbojet
- turbofan

Ram powered jet engine:

- ramjet
- scramjet

Pulsed combustion jet engine:

- pulse detonation engine
- pulse jet engine
- motorjet

## Turbojet Engine

Two engineers, Frank Whittle in the United Kingdom and Hans von Ohain in Germany, developed the turbojet concept independently into practical engines during the late 1930s.

Turbojets consist of an inlet, a compressor, a combustor, a turbine (that drives the compressor) and a propelling nozzle. The compressed air is heated in the combustor and passes through the turbine, then expands in the nozzle to produce a high speed propelling jet

Turbojets have a low propulsive efficiency below about Mach 2 and produce a lot of jet noise, both a result of the very high velocity of the exhaust. Modern jet propelled aircraft are powered by turbofans. These engines, with their lower exhaust velocities, produce less jet noise and use less fuel. Turbojets are still used to power medium range cruise missiles, due to their high exhaust speed, low frontal area and relative simplicity.

## Turbofan Engine

Turbofan engine

Most modern jet engines are turbofans. The low pressure compressor (LPC), usually known as a fan, compresses air into a bypass duct whilst its inner portion supercharges the core compressor. The fan is often an integral part of a multi-stage core LPC. The bypass airflow either passes to a separate 'cold nozzle' or mixes with low pressure turbine exhaust gases, before expanding through a 'mixed flow nozzle'.

In the 1960s there was little difference between civil and military jet engines, apart from the use of afterburning in some (supersonic) applications. Today, turbofans are used for airlinersbecause they have an exhaust speed that is better matched to the subsonic flight speed of the airliner. At airliner flight speeds, the exhaust speed from a turbojet engine is excessively high and wastes energy. The lower exhaust speed from a turbofan gives better fuel consumption. The increased airflow from the fan gives higher thrust at low speeds. The lower exhaust speed also gives much lower jet noise.

Thus civil turbofans today have a low exhaust speed (low *specific thrust* — net thrust divided by airflow) to keep jet noise to a minimum and to improve fuel efficiency. Consequently, the bypass ratio (bypass flow divided by core flow) is relatively high (ratios from 4:1 up to 8:1 are common), with the Rolls-Royce Trent XWB approaching 10:1. Only a single fan stage is required, because a low specific thrust implies a low fan pressure ratio.

Military turbofans, however, have a relatively high specific thrust, to maximize the thrust for a given frontal area, jet noise being of less concern in military uses relative to civil uses. Multistage fans are normally needed to reach the relatively high fan pressure ratio needed for high specific thrust. Although high turbine inlet temperatures are often employed, the bypass ratio tends to be low, usually significantly less than 2.0.

## Major Components

Major components of a turbojet including references to turbofans, turboprops and turboshafts:

## Cold Section

- Air intake (Inlet) — For subsonic aircraft, the inlet is a duct which is required to ensure smooth airflow into the engine despite air approaching the inlet from directions other than straight ahead. This occurs on the ground from cross winds and in flight with aircraft pitch and yaw motions. The duct length is minimised to reduce drag and weight. Air enters the compressor at about half the speed of sound so at flight speeds lower than this the flow will accelerate along the inlet and at higher flight speeds it will slow down. Thus the internal profile of the inlet has to accommodate both accelerating and diffusing flow without undue losses. For supersonic aircraft, the inlet has features such as cones and ramps to produce the most efficient series of shockwaves which form when supersonic flow slows down. The air slows down from the flight speed to subsonic velocity through the shockwaves, then to about half the speed of sound at the compressor through the subsonic part of the inlet. The particular system of shockwaves is chosen, with regard to many constraints such as cost and operational needs, to minimise losses which in turn maximises the pressure recovery at the compressor.
- Compressor or Fan — The compressor is made up of stages. Each stage consists of rotating blades and stationary stators or vanes. As the air moves through the compressor, its pressure and temperature increase. The power to drive the compressor comes from the turbine, as shaft torque and speed.
- Bypass ducts deliver the flow from the fan with minimum losses to the bypass propelling nozzle. Alternatively the fan flow may be mixed with the turbine exhaust before entering a single propelling nozzle. In another arrangement an afterburner may be installed between the mixer and nozzle.
- Shaft — The shaft connects the turbine to the compressor, and runs most of the length of the engine. There may be as many as three concentric shafts, rotating at independent speeds, with as many sets of turbines and compressors. Cooling air for the turbines may flow through the shaft from the compressor.
- Diffuser section: - The diffuser slows down the compressor delivery air to reduce flow losses in the combustor. Slower air is also required to help stabilize the combustion flame and the higher static pressure improves the combustion efficiency.

## Hot Section

- Combustor or Combustion Chamber — Fuel is burned continuously after initially being ignited during the engine start.
- Turbine — The turbine is a series of bladed discs that act like a windmill, extracting energy from the hot gases leaving the combustor. Some of this energy is used to drive the compressor. Turboprop, turboshaft and turbofan engines have additional turbine stages to drive a propeller, bypass fan or helicopter rotor. In a free turbine the turbine driving the compressor

rotates independently of that which powers the propellor or helicopter rotor. Cooling air, bled from the compressor, may be used to cool the turbine blades, vanes and discs to allow higher turbine entry gas temperatures for the same turbine material temperatures.

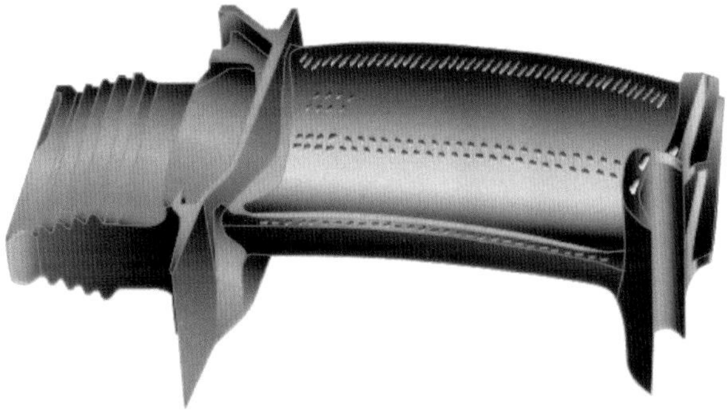

A blade with internal cooling as applied in the high-pressure turbine

- Afterburner or reheat (British) — (mainly military) Produces extra thrust by burning fuel in the jetpipe. This reheating of the turbine exhaust gas raises the propelling nozzle entry temperature and exhaust velocity. The nozzle area is increased to accommodate the higher specific volume of the exhaust gas. This maintains the same airflow through the engine to ensure no change in its operating characteristics.
- Exhaust or Nozzle — Turbine exhaust gases pass through the propelling nozzle to produce a high velocity jet. The nozzle is usually convergent with a fixed flow area.
- Supersonic nozzle — For high nozzle pressure ratios (Nozzle Entry Pressure/Ambient Pressure) a convergent-divergent (de Laval) nozzle is used. The expansion to atmospheric pressure and supersonic gas velocity continues downstream of the throat and produces more thrust.

The various components named above have constraints on how they are put together to generate the most efficiency or performance. The performance and efficiency of an engine can never be taken in isolation; for example fuel/distance efficiency of a supersonic jet engine maximises at about mach 2, whereas the drag for the vehicle carrying it is increasing as a square law and has much extra drag in the transonic region. The highest fuel efficiency for the overall vehicle is thus typically at Mach ~0.85.

For the engine optimisation for its intended use, important here is air intake design, overall size, number of compressor stages (sets of blades), fuel type, number of exhaust stages, metallurgy of components, amount of bypass air used, where the bypass air is introduced, and many other factors. For instance, let us consider design of the air intake.

## Operation

### Engine Cycle

The thermodynamics of a typical air-breathing jet engine are modeled approximately by a Brayton Cycle which is a thermodynamic cycle that describes the workings of the gas turbineengine, which is the basis of the airbreathing jet engine and others. It is named after George Brayton (1830–1892), the American engineer who developed it, although it was originally proposed and patented by Englishman John Barber in 1791. It is also sometimes known as the Joule cycle.

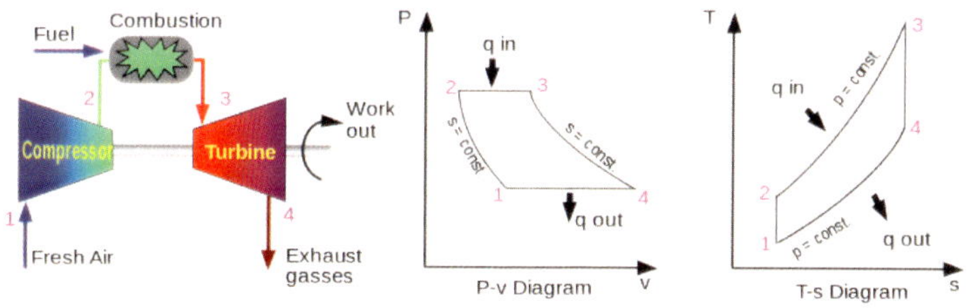


Brayton cycle

## Thrust Lapse

The nominal net thrust quoted for a jet engine usually refers to the Sea Level Static (SLS) condition, either for the International Standard Atmosphere (ISA) or a hot day condition (e.g. ISA+10 °C). As an example, the GE90-76B has a take-off static thrust of 76,000 lbf (360 kN) at SLS, ISA+15 °C.

Naturally, net thrust will decrease with altitude, because of the lower air density. There is also, however, a flight speed effect.

Initially as the aircraft gains speed down the runway, there will be little increase in nozzle pressure and temperature, because the ram rise in the intake is very small. There will also be little change in mass flow. Consequently, nozzle gross thrust initially only increases marginally with flight speed. However, being an air breathing engine (unlike a conventional rocket) there is a penalty for taking on-board air from the atmosphere. This is known as ram drag. Although the penalty is zero at static conditions, it rapidly increases with flight speed causing the net thrust to be eroded.

As flight speed builds up after take-off, the ram rise in the intake starts to have a significant effect upon nozzle pressure/temperature and intake airflow, causing nozzle gross thrust to climb more rapidly. This term now starts to offset the still increasing ram drag, eventually causing net thrust to start to increase. In some engines, the net thrust at say Mach 1.0, sea level can even be slightly greater than the static thrust. Above Mach 1.0, with a subsonic inlet design, shock losses tend to decrease net thrust, however a suitably designed supersonic inlet can give a lower reduction in intake pressure recovery, allowing net thrust to continue to climb in the supersonic regime.

## Safety and Reliability

Jet engines are usually very reliable and have a very good safety record. However, failures do sometimes occur.

## Engine Surge

In some cases in jet engines the conditions in the engine due to airflow entering the engine or other variations can cause the compressor blades to stall. When this occurs the pressure in the engine blows out past the blades, and the stall is maintained until the pressure has decreased, and the engine has lost all thrust. The compressor blades will then usually come out of stall, and re-pressurize the engine. If conditions are not corrected, the cycle will usually repeat. This is called surge. Depending on the engine this can be highly damaging to the engine and creates worrying vibrations for the crew.

## Blade Containment

Fan, compressor or turbine blade failures have to be contained within the engine casing. To do this the engine has to be designed to pass blade containment tests as specified by certification authorities.

## Bird Ingestion

Bird ingestion is the term used when birds enter the intake of a jet engine. It is a common aircraft safety hazard and has caused fatal accidents. In 1988 an Ethiopian Airlines Boeing 737 ingested pigeons into both engines during take-off and then crashed in an attempt to return to the Bahir Dar airport; of the 104 people aboard, 35 died and 21 were injured. In another incident in 1995, a Dassault Falcon 20 crashed at a Paris airport during an emergency landing attempt after ingesting lapwings into an engine, which caused an engine failure and a fire in the airplane fuselage; all 10 people on board were killed. In 2009, an Airbus A320 aircraft, US Airways Flight 1549, ingested one bird into each engine. The plane ditched in the Hudson River after taking off from LaGuardia International Airport in New York City. There were no fatalities.

Jet engines have to be designed to withstand the ingestion of birds of a specified weight and number and to not lose more than a specified amount of thrust. The weight and numbers of birds that can be ingested without hazarding the safe flight of the aircraft are related to the engine intake area. The hazards of ingesting birds beyond the "designed-for" limit were shown on US Airways Flight 1549 when the aircraft struck Canada geese.

The outcome of an ingestion event and whether it causes an accident, be it on a small fast plane, such as military jet fighters, or a large transport, depends on the number and weight of birds and where they strike the fan blade span or the nose cone. Core damage usually results with impacts near the blade root or on the nose cone.

The highest risk of a bird ingestion is during takeoff and landing and during low level flying.

## Volcanic Ash

If a jet plane is flying through air contaminated with volcanic ash, there is risk that ingested ash will cause erosion damage to the compressor blades, blockage of fuel nozzle air holes and blockage of the turbine cooling passages. Some of these effects may cause the engine to surge or flame-out during the flight. Re-lights are usually successful after flame-outs but with considerable loss of altitude. It was the case of British Airways Flight 9 which flew through volcanic dust at 37,000 ft. All 4 engines flamed out and re-light attempts were successful at about 13,000 ft.

## Uncontained Failures

One class of failure that has caused accidents is the uncontained failure, where rotating parts of the engine break off and exit through the case. These high energy parts can cut fuel and control lines, and can penetrate the cabin. Although fuel and control lines are usually duplicated for reliability, the crash of United Airlines Flight 232 was caused when hydraulic fluid lines for all three independent hydraulic systems were simultaneously severed by shrapnel from an uncontained engine failure. Prior to the United 232 crash, the probability of a simultaneous failure of all three hydraulic systems was considered as high as a billion-to-one. However, the statistical models used to come up with this figure did not account for the fact that the number-two engine was mounted at the tail close to all the hydraulic lines, nor the possibility that an engine failure would release many fragments in many directions. Since then, more modern aircraft engine designs have focused on keeping shrapnel from penetrating the cowling or ductwork, and have increasingly utilized high-strength composite materials to achieve the required penetration resistance while keeping the weight low.

## Advanced Designs

### Ramjet

A ramjet is a form of airbreathing jet engine using the engine's forward motion to compress incoming air, without a rotary compressor. Ramjets cannot produce thrust at zero airspeed and thus cannot move an aircraft from a standstill. Ramjets require considerable forward speed to operate well, and as a class work most efficiently at speeds around Mach 3. This type of jet can operate up to speeds of Mach 6.

Ramjets can be particularly useful in applications requiring a small and simple engine for high speed use, such as missiles, while weapon designers are looking to use ramjet technology in artillery shells to give added range: it is anticipated that a 120-mm mortar shell, if assisted by a ramjet, could attain a range of 22 mi (35 km). They have also been used successfully, though not efficiently, as tip jets on helicopter rotors.

Ramjets are frequently confused with pulsejets, which use an intermittent combustion, but ramjets employ a continuous combustion process, and are a quite distinct type of jet engine.

### P&W J58 Mach 3+ afterburning Turbojet

Turbojet operation over the complete flight envelope from zero to Mach 3+ requires features to allow the compressor to function properly at the high inlet temperatures beyond Mach 2.5 as well as at low flight speeds. The J58 compressor solution was to bleed airflow from the 4th compressor stage at speeds above about Mach 2. The bleed flow, 20% at Mach 3, was returned to the engine via 6 external tubes to cool the afterburner liner and primary nozzle as well as to provide extra air for combustion. The J58 engine was the only operational turbojet engine, being designed to operate continuously even at maximum afterburning, for Mach 3.2 cruise.

An alternative solution is seen in a contemporary installation, which did not reach operational status, the Mach 3 GE YJ93/XB-70. It used a variable stator compressor. Yet another solution was specified in a proposal for a Mach 3 reconnaissance Phantom. This was pre-compressor cooling, albeit available for relatively short duration.

### Hydrogen-fuelled air-breathing Jet Engines

Jet engines can be run on almost any fuel. Hydrogen is a highly desirable fuel, as, although the energy per mole is not unusually high, the molecule is very much lighter than other molecules. The energy per kg of hydrogen is twice that of more common fuels and this gives twice the specific impulse. In addition, jet engines running on hydrogen are quite easy to build—the first ever turbojet was run on hydrogen. Also, although not duct engines, hydrogen-fueled rocket engines have seen extensive use.

However, in almost every other way, hydrogen is problematic. The downside of hydrogen is its density; in gaseous form the tanks are impractical for flight, but even in the form of liquid hydrogen it has a density one fourteenth that of water. It is also deeply cryogenic and requires very significant insulation that precludes it being stored in wings. The overall vehicle would end up being very large, and difficult for most airports to accommodate. Finally, pure hydrogen is not found in nature, and must be manufactured either via steam reforming or expensive electrolysis. Nevertheless, research is ongoing and hydrogen-fueled aircraft designs do exist that may be feasible.

## Precooled Jet Engines

An idea originated by Robert P. Carmichael in 1955 is that hydrogen-fueled engines could theoretically have much higher performance than hydrocarbon-fueled engines if a heat exchanger were used to cool the incoming air. The low temperature allows lighter materials to be used, a higher mass-flow through the engines, and permits combustors to inject more fuel without overheating the engine.

This idea leads to plausible designs like Reaction Engines SABRE, that might permit single-stage-to-orbit launch vehicles, and ATREX, which could permit jet engines to be used up to hypersonic speeds and high altitudes for boosters for launch vehicles. The idea is also being researched by the EU for a concept to achieve non-stop antipodal supersonic passenger travel at Mach 5 (Reaction Engines A2).

## Nuclear-powered Ramjet

Project Pluto was a nuclear-powered ramjet, intended for use in a cruise missile. Rather than combusting fuel as in regular jet engines, air was heated using a high-temperature, unshielded nuclear reactor. This dramatically increased the engine burn time, and the ramjet was predicted to be able to cover any required distance at supersonic speeds (Mach 3 at tree-top height).

However, there was no obvious way to stop it once it had taken off, which is a great disadvantage in any non-disposable application. Also, because the reactor was unshielded, it was dangerous to be in or around the flight path of the vehicle (although the exhaust itself wasn't radioactive), and the project was cancelled.

## Scramjets

Scramjets are an evolution of ramjets that are able to operate at much higher speeds than any other kind of airbreathing engine. They share a similar structure with ramjets, being a specially shaped tube that compresses air with no moving parts through ram-air compression. Scramjets, however, operate with supersonic airflow through the entire engine. Thus, scramjets do not have the diffuser required by ramjets to slow the incoming airflow to subsonic speeds.

Scramjets start working at speeds of at least Mach 4, and have a maximum useful speed of approximately Mach 17. Due to aerodynamic heating at these high speeds, cooling poses a challenge to engineers.

## Turborocket

The air turborocket is a form of combined-cycle jet engine. The basic layout includes a gas generator, which produces high pressure gas, that drives a turbine/compressor assembly which compresses atmospheric air into a combustion chamber. This mixture is then combusted before leaving the device through a nozzle and creating thrust.

There are many different types of air turborockets. The various types generally differ in how the gas generator section of the engine functions.

Air turborockets are often referred to as turboramjets, turboramjet rockets, turborocket expanders, and many others. As there is no consensus on which names apply to which specific concepts, various sources may use the same name for two different concepts.

## Terminology

To specify the RPM, or rotor speeds, of a jet engine, abbreviations are commonly used:

- For a turboprop engine, $N_p$ refers to the RPM of the propeller shaft. For example, a common $N_p$ would be about 2200 RPM for a constant speed propeller.
- $N_1$ or $N_g$ refers to the RPM of the gas generator section. Each engine manufacturer will pick between those two abbreviation. N1 is also used for the fan speed on a turbofan, in which case $N_2$ is the gas generator speed (2 shaft engine). Ng is mainly used for turboprop or turboshaft engines. For example, a common $N_g$ would be on the order of 30,000 RPM.
- $N_2$ or $N_f$ refers to the speed of the power turbine section. Each engine manufacturer will pick between those two abbreviations but N2 is mainly used for turbofan engines whereas Nf is mainly used for turboprop or turboshaft engines. In many cases, even for free turbine engines, the $N_1$ and $N_2$ may be very similar.
- $N_s$ refers to the speed of the reduction gear box (RGB) output shaft for turboshaft engines.

In many cases, instead of expressing rotor speeds ($N_1$, $N_2$) as RPM on cockpit displays, pilots are provided with the speeds expressed as a percentage of the design point speed. For example, at full power, the $N_1$ might be 101.5% or 100%. This user interface decision has been made as a human factors consideration, since pilots are more likely to notice a problem with a two- or 3-digit percentage (where 100% implies a nominal value) than with a 5-digit RPM.

## Turbofan vs Turbojet

It seems unbelievable now – but once upon a time, commercial aircraft and private jets were fueled by loud, fuel-guzzling turbojet engines. We've come a long way since the World War II era. Today's turbofan-driven aircraft are quiet and fuel-efficient.

All private jet engines operate differently – but most run by forcing air into a tube where it's then compressed, mixed with fuel, burned, and exhausted at a very high speed in order to produce thrust. This basic operating principle is true of both turbojet and turbofan jet engines.

## Early Days: The Turbojet Engine

How does it work? The turbojet engine compresses air in a chamber, heats it, and lets it expand through a turbine and into a nozzle where it then accelerates to a high rate of speed. The concept itself is fairly simple; in fact, numerous aircraft during World War II employed this principle. Unfortunately, the rudimentary design and fast exhaust speed of the turbojet mean the engine is really too loud and not efficient enough for private jet charters.

## Today's Choice: The Turbofan Engine

On the other hand, the turbofan engine is ideal because it possesses high thrust and fuel efficiency capabilities. Turbofans run basically the same way turbojets do – however, they incorporate a fan component. Air comes in the engine through an inlet and passes through the fan; the rest travels to the burner to be mixed with fuel and ignited to generate propulsion. In contrast with turbojets, turbofans actually get their thrust from two sources: the fan and the engine core.

The result? Turbofans have much less exhaust than turbofans, and they are much more efficient at subsonic speeds. Because the majority of charter jet flights operate at subsonic speeds, the turbofan is a better engine choice for a charter aircraft. In an aircraft with a turbofan engine, charter pilots can pull back the throttle to cruise, which saves hundreds of dollars on fuel during a flight. Knowing all this, it's no surprise that the turbofan is the clear engine of choice for both commercial airlines and private jet charters.

## References


- Wake, M.H. (1993). "The Skull as a Locomotor Organ". In Hanken, James. The Skull. University of Chicago Press. p. 460. ISBN 978-0-226-31573-7
- Madin, L. P. (1990). "Aspects of jet propulsion in salps". Canadian Journal of Zoology. 68: 765–777. doi:10.1139/z90-111
- Winter, Frank H. (December 1980). "Ducted Fan or the World's First Jet Plane? The Coanda claim re-examined". The Aeronautical Journal. Royal Aeronautical Society. 84
- "ASH 26 E Information". DE: Alexander Schleicher. Archived from the original on 2006-10-08. Retrieved 2006-11-24
- Chamberlain Jr, John A. (1987). "32. Locomotion of Nautilus". In Saunders, W. B.; Landman, N. H. Nautilus: The Biology and Paleobiology of a Living Fossil. ISBN 9789048132980
- Packard, A. (1972). "Cephalopods and Fish: the Limits of Convergence". Biological Reviews. 47 (2): 241–307. doi:10.1111/j.1469-185X.1972.tb00975.x
- "Water injection pre-compressor cooling assist space access"Mehta, Huynh, Hagseth, The Aeronautical Journal, February 2015, Volume 19, Number 1212, p.147
- "English Electric Lightning F53 (53-671) – Power Plants". Gatwick Aviation Museum. Retrieved 9 June 2018
- Ian McNeil, ed. (1990). Encyclopedia of the History of Technology. London: Routledge. pp. 315–21. ISBN 0-203-19211-7
- Sutherland, K. R.; Madin, L. P. (2010). "Comparative jet wake structure and swimming performance of salps". Journal of Experimental Biology. 213 (Pt 17): 2967–75. doi:10.1242/jeb.041962. PMID 20709925
- "Gas Turbine Technology Evolution: A Designer's Perspective" Bernard L.Koff Journal of Propulsion and Power Vol20 No4 July–August 2004 Fig.34/41

# Aircraft Control Systems

4

An aircraft flight control system consists of flight control surfaces, cockpit controls and the necessary operating mechanisms to control the direction of the flight. The topics elaborated in this chapter will help in presenting a better perspective of flight control surfaces and aircraft engine controls.

## Aircraft Flight Control System

As aviation matured and aircraft designers learned more about aerodynamics, the industry produced larger and faster aircraft. Therefore, the aerodynamic forces acting upon the control surfaces increased exponentially. To make the control force required by pilots manageable, aircraft engineers designed more complex systems. At first, hydromechanical designs, consisting of a mechanical circuit and a hydraulic circuit, were used to reduce the complexity, weight, and limitations of mechanical flight controls systems.

As aircraft became more sophisticated, the control surfaces were actuated by electric motors, digital computers, or fiber optic cables. Called "fly-by-wire," this flight control system replaces the physical connection between pilot controls and the flight control surfaces with an electrical interface. In addition, in some large and fast aircraft, controls are boosted by hydraulically or electrically actuated systems. In both the fly-by-wire and boosted controls, the feel of the control reaction is fed back to the pilot by simulated means.

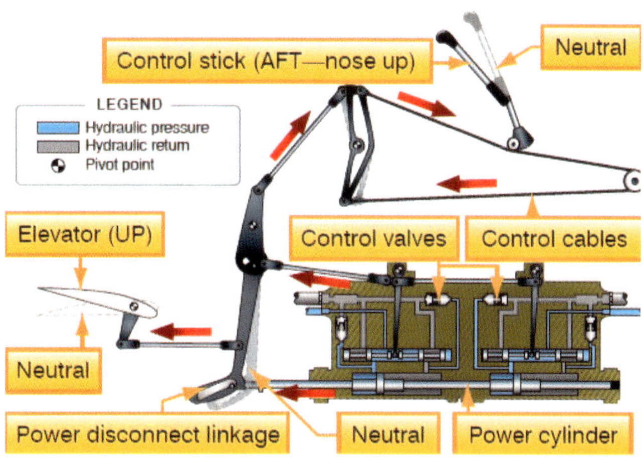

Hydromechanical flight control system.

Current research at the National Aeronautics and Space Administration (NASA) Dryden Flight Research Center involves Intelligent Flight Control Systems (IFCS). The goal of this project is to develop an adaptive neural network-based flight control system. Applied directly to flight control system feedback errors, IFCS provides adjustments to improve aircraft performance in normal flight, as well as with system failures. With IFCS, a pilot is able to maintain control and safely land an aircraft that has suffered a failure to a control surface or damage to the airframe. It also improves mission capability, increases the reliability and safety of flight, and eases the pilot workload.

Today's aircraft employ a variety of flight control systems. For example, some aircraft in the sport pilot category rely on weight-shift control to fly while balloons use a standard burn technique. Helicopters utilize a cyclic to tilt the rotor in the desired direction along with a collective to manipulate rotor pitch and anti-torque pedals to control yaw.

## Description

Aircraft flight controls are the means by which a pilot controls the direction and attitude of an aircraft in flight.

Flight control systems are subdivided into what are referred to as primary and secondary flight controls. Primary flight controls are required to safely control an aircraft during flight and consist of ailerons, elevators (or, in some installations, stabilator) and rudder. Secondary flight controls are intended to improve the aircraft performance characteristics or to relieve excessive control loading, and consist of high lift devices such as slats and flaps as well as flight spoilers and trim systems.

## Cockpit controls

### Primary Controls

Generally, the primary cockpit flight controls are arranged as follows:

- a control yoke (also known as a control column), centre stick or side-stick (the latter two also colloquially known as a control or joystick), governs the aircraft's roll and pitch by moving the ailerons (or activating wing warping on some very early aircraft designs) when turned or deflected left and right, and moves the elevators when moved backwards or forwards
- rudder pedals, or the earlier, pre-1919 "rudder bar", to control yaw, which move the rudder; left foot forward will move the rudder left for instance.
- throttle controls to control engine speed or thrust for powered aircraft.

The control yokes also vary greatly amongst aircraft. There are yokes where roll is controlled by rotating the yoke clockwise/counterclockwise (like steering a car) and pitch is controlled by tilting the control column towards you or away from you, but in others the pitch is controlled by sliding the yoke into and out of the instrument panel (like most Cessnas, such as the 152 and 172), and in some the roll is controlled by sliding the whole yoke to the left and right (like the Cessna 162). Centre sticks also vary between aircraft. Some are directly connected to the control surfaces using cables, others (fly-by-wire airplanes) have a computer in between which then controls the electrical actuators.

Even when an aircraft uses variant flight control surfaces such as a V-tail ruddervator, flaperons, or elevons, to avoid pilot confusion the aircraft's flight control system will still be designed so that the stick or yoke controls pitch and roll conventionally, as will the rudder pedals for yaw. The basic pattern for modern flight controls was pioneered by French aviation figure Robert Esnault-Pelterie, with fellow French aviator Louis Blériot popularizing Esnault-Pelterie's control format initially on Louis' Blériot VIII monoplane in April 1908, and standardizing the format on the July 1909 Channel-crossing Blériot XI. Flight control has long been taught in such fashion for many decades, as popularized in ab initio instructional books such as the 1944 work Stick and Rudder.

In some aircraft, the control surfaces are not manipulated with a linkage. In ultralight aircraft and motorized hang gliders, for example, there is no mechanism at all. Instead, the pilot just grabs the lifting surface by hand (using a rigid frame that hangs from its underside) and moves it.

## Secondary Controls

In addition to the primary flight controls for roll, pitch, and yaw, there are often secondary controls available to give the pilot finer control over flight or to ease the workload. The most commonly available control is a wheel or other device to control elevator trim, so that the pilot does not have to maintain constant backward or forward pressure to hold a specific pitch attitude (other types of trim, for rudder and ailerons, are common on larger aircraft but may also appear on smaller ones). Many aircraft have wing flaps, controlled by a switch or a mechanical lever or in some cases are fully automatic by computer control, which alter the shape of the wing for improved control at the slower speeds used for take-off and landing. Other secondary flight control systems may be available, including slats, spoilers, air brakes and variable-sweep wings.

## Flight Control Systems

## Mechanical

de Havilland Tiger Moth elevator and rudder cables

Mechanical or manually operated flight control systems are the most basic method of controlling an aircraft. They were used in early aircraft and are currently used in small aircraft where the aerodynamic forces are not excessive. Very early aircraft, such as the Wright Flyer I, Blériot XI and Fokker Eindecker used a system of wing warping where no conventionally hinged control surfaces were used on the wing, and sometimes not even for pitch control as on the Wright Flyer I and original versions of the 1909 Etrich Taube, which only had a hinged/pivoting rudder in addition to the warping-operated pitch and roll controls. A manual flight control system uses a collection of mechanical parts such as pushrods, tension cables, pulleys, counterweights, and sometimes chains to transmit the forces applied to the cockpit controls directly to the control surfaces. Turnbuckles are often used to adjust control cable tension. The Cessna Skyhawk is a typical example of an aircraft that uses this type of system. Gust locks are often used on parked aircraft with mechanical systems to protect the control surfaces and linkages from damage from wind. Some aircraft have gust locks fitted as part of the control system.

Increases in the control surface area required by large aircraft or higher loads caused by high airspeeds in small aircraft lead to a large increase in the forces needed to move them, consequently complicated mechanical gearing arrangements were developed to extract maximum mechanical advantage in order to reduce the forces required from the pilots. This arrangement can be found on bigger or higher performance propeller aircraft such as the Fokker 50.

Some mechanical flight control systems use servo tabs that provide aerodynamic assistance. Servo tabs are small surfaces hinged to the control surfaces. The flight control mechanisms move these tabs, aerodynamic forces in turn move, or assist the movement of the control surfaces reducing the amount of mechanical forces needed. This arrangement was used in early piston-engined transport aircraft and in early jet transports. The Boeing 737 incorporates a system, whereby in the unlikely event of total hydraulic system failure, it automatically and seamlessly reverts to being controlled via servo-tab.

## Hydro-mechanical

The complexity and weight of mechanical flight control systems increase considerably with the size and performance of the aircraft. Hydraulically powered control surfaces help to overcome these limitations. With hydraulic flight control systems, the aircraft's size and performance are limited by economics rather than a pilot's muscular strength. At first, only-partially boosted systems were used in which the pilot could still feel some of the aerodynamic loads on the control surfaces.

A hydro-mechanical flight control system has two parts:

- The *mechanical circuit*, which links the cockpit controls with the hydraulic circuits. Like the mechanical flight control system, it consists of rods, cables, pulleys, and sometimes chains.
- The *hydraulic circuit*, which has hydraulic pumps, reservoirs, filters, pipes, valves and actuators. The actuators are powered by the hydraulic pressure generated by the pumps in the hydraulic circuit. The actuators convert hydraulic pressure into control surface movements. The electro-hydraulic servo valves control the movement of the actuators.

The pilot's movement of a control causes the mechanical circuit to open the matching servo valve in the hydraulic circuit. The hydraulic circuit powers the actuators which then move the control surfaces. As the actuator moves, the servo valve is closed by a mechanical feedback linkage - one that stops movement of the control surface at the desired position.

This arrangement was found in the older-designed jet transports and in some high-performance aircraft. Examples include the Antonov An-225 and the Lockheed SR-71.

### Artificial Feel Devices

With purely mechanical flight control systems, the aerodynamic forces on the control surfaces are transmitted through the mechanisms and are felt directly by the pilot, allowing tactile feedback of airspeed. With hydromechanical flight control systems, however, the load on the surfaces cannot be felt and there is a risk of overstressing the aircraft through excessive control surface movement. To overcome this problem, artificial feel systems can be used. For example, for the controls of the RAF's Avro Vulcan jet bomber and the RCAF's Avro Canada CF-105 Arrow supersonic interceptor (both 1950s-era designs), the required force feedback was achieved by a spring device. The fulcrumof this device was moved in proportion to the square of the air speed (for the elevators) to give increased resistance at higher speeds. For the controls of the American Vought F-8 Crusader and the LTV A-7 Corsair II warplanes, a 'bob-weight' was used in the pitch axis of the control stick, giving force feedback that was proportional to the airplane's normal acceleration.

### Stick Shaker

A stick shaker is a device (available in some hydraulic aircraft) that is attached to the control column, which shakes the control column when the aircraft is about to stall. Also in some aircraft like the McDonnell Douglas DC-10 there is/was a back-up electrical power supply that the pilot can turn on to re-activate the stick shaker in case the hydraulic connection to the stick shaker is lost.

### Fly-by-wire Control Systems

A fly-by-wire (FBW) system replaces manual flight control of an aircraft with an electronic interface. The movements of flight controls are converted to electronic signals transmitted by wires (hence the fly-by-wire term), and flight control computers determine how to move the actuators at each control surface to provide the expected response. Commands from the computers are also input without the pilot's knowledge to stabilize the aircraft and perform other tasks. Electronics for aircraft flight control systems are part of the field known as avionics.

Fly-by-optics, also known as *fly-by-light*, is a further development using fiber optic cables.

## Flight Control Surfaces

The mechanics of flight are highly complex, encompassing principles such as the four forces and axes, as well as technical terms such as control surfaces, adverse yaw and coordinated flight. Wind-axis systems, which refer to aircraft forces in relation to the direction of the aircraft's velocity, are not discussed here to avoid confusion. Also, some of the explanations given are highly simplified in order to allow educators to help students visualize the principle being discussed. For educators with an aviation background and capable students, modification of these activities and background information is encouraged.

## The Forces of Flight

Every aircraft, whether an airplane, helicopter or rocket, is affected by four opposing forces: Thrust, Lift, Drag and Weight. Control surfaces, such as the rudder or ailerons, adjust the direction of these forces, allowing the pilot to use them in the most advantageous way possible.

A force can be thought of as a push or pull in a specific direction. It is a vector quantity, which means a force has both a magnitude (amount) and a direction.

## Controlling the Motion of Flight

In order for an aircraft to reach its destination, the forces of flight have to be precisely manipulated. To do this, the aircraft has control surfaces which can direct airflow in very specific ways.

## Flight Control Surfaces

The primary flight control surfaces on a fixed-wing aircraft include: ailerons, elevators, and the rudder. The ailerons are attached to the trailing edge of both wings and when moved, rotate the aircraft around the longitudinal axis. The elevator is attached to the trailing edge of the horizontal stabilizer. When it is moved, it alters aircraft pitch, which is the attitude about the horizontal or lateral axis. The rudder is hinged to the trailing edge of the vertical stabilizer. When the rudder changes position, the aircraft rotates about the vertical axis (yaw). The primary flight controls of a light aircraft and the movement they create relative to the three axes of flight.

## Development

The Wright brothers are credited with developing the first practical control surfaces. It is a main part of their patent on flying. Unlike modern control surfaces, they used wing warping. In an attempt to circumvent the Wright patent, Glenn Curtiss made hinged control surfaces, the same type of concept first patented some four decades earlier in the United Kingdom. Hinged control surfaces have the advantage of not causing stresses that are a problem of wing warping and are easier to build into structures.

## Axes of Motion

An aircraft is free to rotate around three axes that are perpendicular to each other and intersect at its center of gravity (CG). To control position and direction a pilot must be able to control rotation about each of them.

## Transverse Axis

The *transverse axis*, also known as *lateral axis*, passes through an aircraft from wingtip to wingtip. Rotation about this axis is called *pitch*. Pitch changes the vertical direction that the aircraft's nose is pointing. The elevators are the primary control surfaces for pitch.

## Longitudinal Axis

The longitudinal axis passes through the aircraft from nose to tail. Rotation about this axis is called *roll*. The angular displacement about this axis is called bank. The pilot changes bank angle by increasing the lift on one wing and decreasing it on the other. This differential lift causes rotation around the longitudinal axis. The ailerons are the primary control of bank. The rudder also has a secondary effect on bank.

## Vertical Axis

The vertical axis passes through an aircraft from top to bottom. Rotation about this axis is called *yaw*. Yaw changes the direction the aircraft's nose is pointing, left or right. The primary control of yaw is with the rudder. Ailerons also have a secondary effect on yaw.

It is important to note that these axes move with the aircraft, and change relative to the earth as the aircraft moves. For example, for an aircraft whose left wing is pointing straight down, its "vertical" axis is parallel with the ground, while its "transverse" axis is perpendicular to the ground.

## Main Control Surfaces

The main control surfaces of a fixed-wing aircraft are attached to the airframe on hinges or tracks so they may move and thus deflect the air stream passing over them. This redirection of the air stream generates an unbalanced force to rotate the plane about the associated axis.

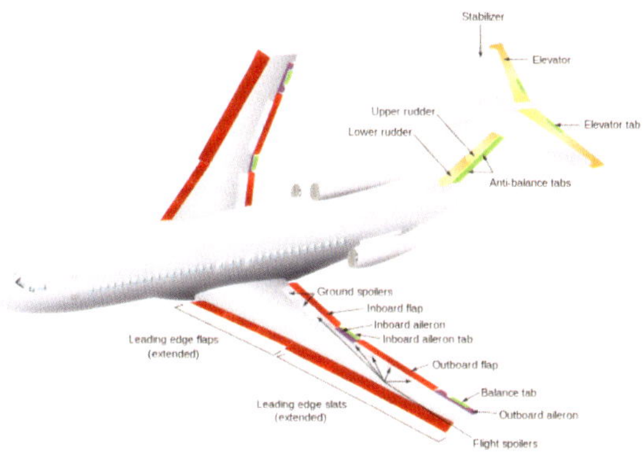


Flight control surfaces of Boeing 727

## Ailerons

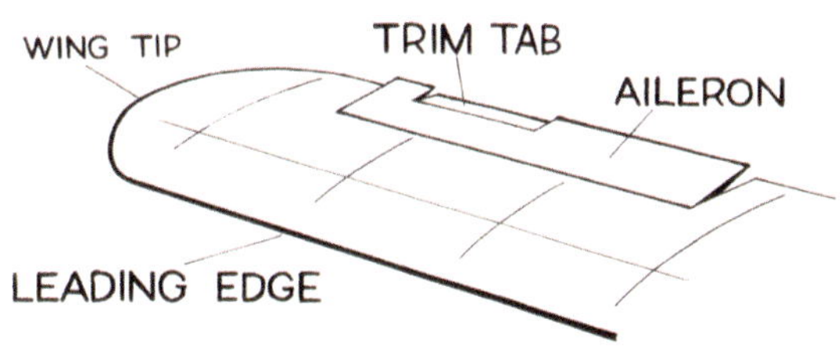


Aileron surface

Ailerons are mounted on the trailing edge of each wing near the wingtips and move in opposite directions. When the pilot moves the stick left, or turns the wheel counter-clockwise, the left aileron goes up and the right aileron goes down. A raised aileron reduces lift on that wing and a lowered one increases lift, so moving the stick left causes the left wing to drop and the right wing to rise. This causes the aircraft to roll to the left and begin to turn to the left. Centering the stick returns the ailerons to neutral maintaining the bank angle. The aircraft will continue to turn until opposite aileron motion returns the bank angle to zero to fly straight.

### Elevator

The elevator is a moveable part of the horizontal stabilizer, hinged to the back of the fixed part of the horizontal tail. The elevators move up and down together. When the pilot pulls the stick backward, the elevators go up. Pushing the stick forward causes the elevators to go down. Raised elevators push down on the tail and cause the nose to pitch up. This makes the wings fly at a higher angle of attack, which generates more lift and more drag. Centering the stick returns the elevators to neutral and stops the change of pitch. Some aircraft, such as an MD-80, use a servo tab within the elevator surface to aerodynamically move the main surface into position. The direction of travel of the control tab will thus be in a direction opposite to the main control surface. It is for this reason that an MD-80 tail looks like it has a 'split' elevator system.

In the canard arrangement, the elevators are hinged to the rear of a foreplane and move in the opposite sense, for example when the pilot pulls the stick back the elevators go down to increase the lift at the front and lift the nose up.

### Rudder

The rudder is typically mounted on the trailing edge of the vertical stabilizer, part of the empennage. When the pilot pushes the left pedal, the rudder deflects left. Pushing the right pedal causes the rudder to deflect right. Deflecting the rudder right pushes the tail left and causes the nose to yaw to the right. Centering the rudder pedals returns the rudder to neutral and stops the yaw.

## Secondary Effects of Controls

### Ailerons

The ailerons primarily control roll. Whenever lift is increased, induced drag is also increased. When the stick is moved left to roll the aircraft to the left, the right aileron is lowered which increases lift on the right wing and therefore increases induced drag on the right wing. Using ailerons causes adverse yaw, meaning the nose of the aircraft yaws in a direction opposite to the aileron application. When moving the stick to the left to bank the wings, adverse yaw moves the nose of the aircraft to the *right*. Adverse yaw is more pronounced for light aircraft with long wings, such as gliders. It is counteracted by the pilot with the rudder. Differential ailerons are ailerons which have been rigged such that the downgoing aileron deflects less than the upward-moving one, reducing adverse yaw.

### Rudder

The rudder is a fundamental control surface which is typically controlled by pedals rather than at the stick. It is the primary means of controlling yaw—the rotation of an airplane about its vertical axis. The rudder may also be called upon to counter-act the adverse yaw produced by the roll-control surfaces.

If rudder is continuously applied in level flight the aircraft will yaw initially in the direction of the applied rudder – the primary effect of rudder. After a few seconds the aircraft will tend to bank in the direction of yaw.

This arises initially from the increased speed of the wing opposite to the direction of yaw and the reduced speed of the other wing. The faster wing generates more lift and so rises, while the other wing tends to go down because of generating less lift. Continued application of rudder sustains rolling tendency because the aircraft flying at an angle to the airflow - skidding towards the forward wing. When applying right rudder in an aircraft with dihedral the left hand wing will have increased angle of attack and the right hand wing will have decreased angle of attack which will result in a roll to the right. An aircraft with anhedral will show the opposite effect. This effect of the rudder is commonly used in model aircraft where if sufficient diheral or polyhedral is included in the wing design, primary roll control such as ailerons may be omitted altogether.

## Turning the Aircraft

Unlike turning a boat, changing the direction of an aircraft normally must be done with the ailerons rather than the rudder. The rudder turns (yaws) the aircraft but has little effect on its direction of travel. With aircraft, the change in direction is caused by the horizontal component of lift, acting on the wings. The pilot tilts the lift force, which is perpendicular to the wings, in the direction of the intended turn by rolling the aircraft into the turn. As the bank angle is increased, the lifting force can be split into two components: one acting vertically and one acting horizontally.

If the total lift is kept constant, the vertical component of lift will decrease. As the weight of the aircraft is unchanged, this would result in the aircraft descending if not countered. To maintain level flight requires increased positive (up) elevator to increase the angle of attack, increase the total lift generated and keep the vertical component of lift equal with the weight of the aircraft. This cannot continue indefinitely. The total load factor required to maintain level flight is directly related to the bank angle. This means that for a given airspeed, level flight can only be maintained up to a certain given angle of bank. Beyond this angle of bank, the aircraft will suffer an accelerated stall if the pilot attempts to generate enough lift to maintain level flight.

## Alternate main Control Surfaces

Some aircraft configurations have non-standard primary controls. For example, instead of elevators at the back of the stabilizers, the entire tailplane may change angle. Some aircraft have a tail in the shape of a V, and the moving parts at the back of those combine the functions of elevators and rudder. Delta wing aircraft may have "elevons" at the back of the wing, which combine the functions of elevators and ailerons.

## Secondary Control Surfaces

### Spoilers

On low drag aircraft such as sailplanes, spoilers are used to disrupt airflow over the wing and greatly reduce lift. This allows a glider pilot to lose altitude without gaining excessive airspeed. Spoilers are sometimes called "lift dumpers". Spoilers that can be used asymmetrically are called spoilerons and can affect an aircraft's roll.

KLM Fokker 70, showing position of flap and liftdumpers flight controls. The liftdumpers are the lifted cream-coloured panels on the wing upper surface (in this picture there are five on the right wing). The flaps are the large drooped surfaces on the trailing edge of the wing.

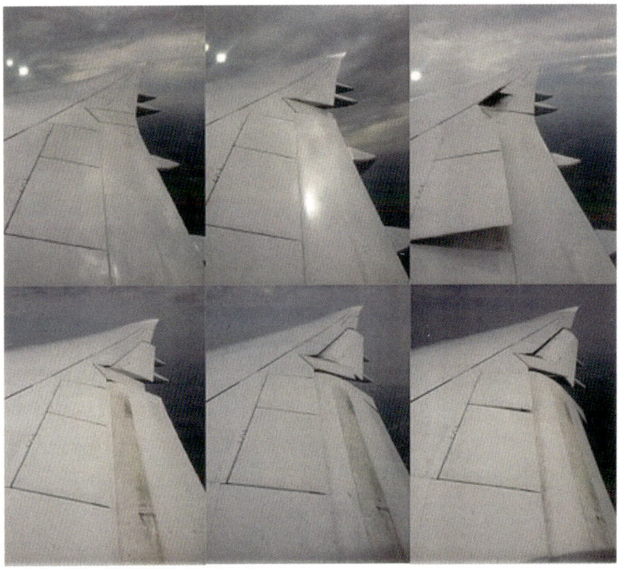

Wing trailing edge flight control surfaces of a Boeing_747-8. Top left: All surfaces at neutral position; Top middle: Right aileron is lowered; Top right: spoilers raised during flight; Middle row: Fowler flaps extended (left), extended more (middle), hinged with inboard slotted part hinged even more (right);

## Flaps

Flaps are mounted on the trailing edge on the inboard section of each wing (near the wing roots). They are deflected down to increase the effective curvature of the wing. Flaps raise the maximum lift coefficient of the aircraft and therefore reduce its stalling speed. They are used during low speed, high angle of attack flight including take-off and descent for landing. Some aircraft are equipped with "flaperons", which are more commonly called "inboard ailerons". These devices function primarily as ailerons, but on some aircraft, will "droop" when the flaps are deployed, thus acting as both a flap and a roll-control inboard aileron.

## Slats

Slats, also known as *leading edge devices*, are extensions to the front of a wing for lift augmentation, and are intended to reduce the stalling speed by altering the airflow over the wing. Slats may be fixed or retractable - fixed slats (e.g. as on the Fieseler Fi 156 Storch) give excellent slow speed and STOL capabilities, but compromise higher speed performance. Retractable slats, as seen on most airliners, provide reduced stalling speed for take-off and landing, but are retracted for cruising.

## Air Brakes

Air brakes on the rear fuselage of a Eurowings BAe 146-300

Air brakes are used to increase drag. Spoilers might act as air brakes, but are not pure air brakes as they also function as lift-dumpers or in some cases as roll control surfaces. Air brakes are usually surfaces that deflect outwards from the fuselage (in most cases symmetrically on opposing sides) into the airstream in order to increase form-drag. As they are in most cases located elsewhere on the aircraft, they do not directly affect the lift generated by the wing. Their purpose is to slow down the aircraft. They are particularly useful when a high rate of descent is required. They are common on high performance military aircraft as well as civilian aircraft, especially those lacking reverse thrust capability.

## Control Trimming Surfaces

Trimming controls allow a pilot to balance the lift and drag being produced by the wings and control surfaces over a wide range of load and airspeed. This reduces the effort required to adjust or maintain a desired flight attitude.

## Elevator Trim

Elevator trim balances the control force necessary to maintain the correct aerodynamic force on the tail to balance the aircraft. Whilst carrying out certain flight exercises, a lot of trim could be required to maintain the desired angle of attack. This mainly applies to slow flight, where a nose-up attitude is required, in turn requiring a lot of trim causing the tailplane to exert a strong downforce. Elevator trim is correlated with the speed of the airflow over the tail, thus airspeed changes to the aircraft require re-trimming. An important design parameter for aircraft is the stability of the aircraft when trimmed for level flight. Any disturbances such as gusts or turbulence will be damped over a short period of time and the aircraft will return to its level flight trimmed airspeed.

## Trimming Tail Plane

Except for very light aircraft, trim tabs on the elevators are unable to provide the force and range of motion desired. To provide the appropriate trim force the entire horizontal tail plane is made adjustable in pitch. This allows the pilot to select exactly the right amount of positive or negative lift from the tail plane while reducing drag from the elevators.

### Control Horn

A control horn is a section of control surface which projects ahead of the pivot point. It generates a force which tends to increase the surface's deflection thus reducing the control pressure experienced by the pilot. Control horns may also incorporate a counterweight which helps to balance the control and prevent it from "fluttering" in the airstream. Some designs feature separate anti-flutter weights.

(In radio controlled model aircraft, the term "control horn" has a different meaning.)

### Spring Trim

In the simplest arrangement trimming is done by a mechanical spring (or bungee) which adds appropriate force to augment the pilot's control input. The spring is usually connected to an elevator trim lever to allow the pilot to set the spring force applied.

### Rudder and Aileron Trim

Trim often does not only apply to the elevator, as there is also trim for the rudder and ailerons in larger aircraft. The use of this is to counter the effects of slip stream, or to counter the effects of the centre of gravity being to one side. This can be caused by a larger weight on one side of the aircraft compared to the other, such as when one fuel tank has a lot more fuel in it than the other.

### Yoke (aeronautics)

The control wheel of an aircraft, akin to a automobile steering wheel.

It's the traditional form of the joystick used as leverage to control the attitude of the plane. Yokes work by mechanical-driven system in which they are directly connected to the control surfaces with cables and rods. They require extra effort as the arm movement gives a longer lever to operate since yokes are larger in size.

Most yokes are coupled together, having a single track of mind when one makes an instant control input unlike the stick which can make independent, multiple instructions to the aircraft.

Aside from giving visual feedback to control positions, they are also less sensitive to maneuver as compared to sticks because they take up a lot of space, concealing other controls from the their line of view.

Yokes are popular among all types of aircraft, for example, the GA aircraft with a conventional control system like the Cessna 172 to the fly-by-wire airliner like the Boeing 787.

### Styles

Yokes come in a variety of shapes and sizes, the most common being of a "U" or "W" design. Some aircraft use an "M" style, such as Embraer aircraft and the Concorde. There are some rarer exotic or archaic styles, such as circular designs much like a steering wheel.

"W"/"U" style yoke in a Cessna 152 light aircraft, mounted on a horizontal tube protruding from the instrumental panel.

In larger aircraft they are usually mounted on a post protruding vertically from the floor, referred to as a control column. In most other planes, they are mounted on a horizontal tube that comes out of the instrument panel.

In the case of the Cirrus SR20 and Cirrus SR22, although the control looks like a side stick, it works like a yoke handle (referred to in the industry as a "side yoke"). The Cessna 162 uses a slightly similar approach.

## Advantages and Disadvantages

Side-sticks and centre-sticks are better for making rapid control inputs and dealing with high g-forces, hence their use in military, sport, and aerobatic aircraft. However, yokes are less sensitive (i.e., more precise) thanks to a larger range of motion and provide more visual feedback to the pilot.

Most yokes are connected and will both move together, thus providing instant indication to the other pilot when one makes a control input. This is in contrast to some fly-by-wire control sticks that allow each pilot to send different, and sometimes greatly conflicting, inputs; these have been known to be responsible for the crash of an Airbus A330 in 2009.

Yokes take up more room than side-sticks in the cockpit, and may even obscure some instruments; by comparison, side-sticks have minimal cockpit intrusion, allowing for the inclusion of retractable tray-tables and making it easier to enter/leave small cockpits.

A yoke, unlike a side-stick, may be used comfortably with either hand. This can be useful if one needs to write or manipulate other controls in the cockpit. This advantage is shared with the centre-stick.

## Centre Stick

A centre stick (or center stick in the United States), or simply control stick is an aircraft cockpit arrangement where the control column (or joystick) is located in the center of the cockpit between the pilot's legs. Since the throttle controls are typically located to the left of the pilot, the right hand is used for the stick, although left-hand or both-hands operation is possible if required.

A fly-by-wire centre stick in a Eurofighter Typhoon cockpit.

Central forward area of the Mirage III cockpit, showing a centre stick.

The centre stick is a part of an aircraft's flight control system, and is typically linked to its ailerons and elevators, or alternatively to its elevons, by control rods or control cables on basic aircraft. On heavier, faster, more advanced aircraft the centre stick may also control power-assist modules. Modern aircraft centre sticks are also usually equipped with a number of electrical control switches within easy finger reach, in order to reduce the pilot's workload.

The centre stick is used in many military fighter jets such as Eurofighter Typhoon and the Mirage III, but also in light aircraft such as Piper Cubs and the Diamond Aircraft line of products such as the DA20, DA40 and DA42.

This arrangement contrasts with the more recently developed "side-stick" which is used in such military fighter jets as the F-16, the F-35 Lightning II and Rafale and also on civil aircraft such as the Airbus A320.

## Aileron

Ailerons are a primary flight control surface which control movement about the longitudinal axis of an aircraft. This movement is referred to as "roll". The ailerons are attached to the outboard trailing edge of each wing and, when a manual or autopilot control input is made, move in opposite directions from one another. In some large aircraft, two ailerons are mounted on each wing. In this configuration, both ailerons on each wing are active during slow speed flight. However, at higher speed, the outboard aileron is locked and only the inboard or high speed aileron is functional.

The ailerons are used to bank the aircraft; to cause one wing tip to move up and the other wing tip to move down. The banking creates an unbalanced side force component of the large wing lift force which causes the aircraft's flight path to curve. (Airplanes turn because of banking created by the ailerons, not because of a rudder input.

The ailerons work by changing the effective shape of the airfoil of the outer portion of the wing. changing the angle of deflection at the rear of an airfoil will change the amount of lift generated by the foil. With greater downward deflection, the lift will increase in the upward direction.

The lift force (Fr or Fl) of the wing section through the aileron is applied at the aerodynamic center of the section which is some distance (L) from the aircraft center of gravity. This creates a torque

$$T = F * L$$

about the center of gravity. If the forces (and distances) are equal there is no net torque on the aircraft. But if the forces are unequal, there is a net torque and the aircraft rotates about its center of gravity. For the conditions shown in the figure, the resulting motion will roll the aircraft to the right (clockwise) as viewed from the rear. If the pilot reverses the aileron deflections (right aileron down, left aileron up) the aircraft will roll in the opposite direction.

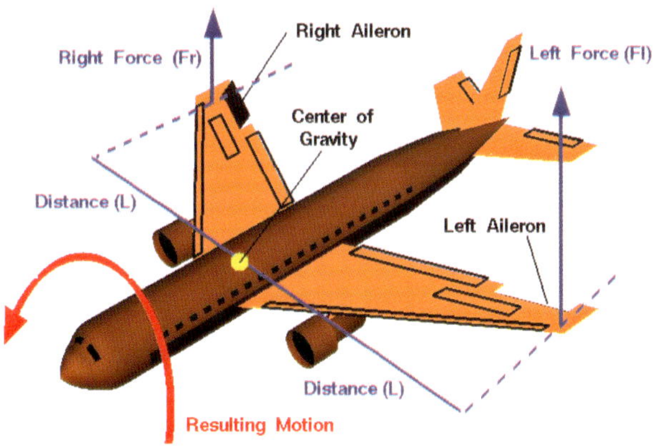


Ailerons

## Flight Dynamics

Pairs of ailerons are typically interconnected so that when one is moved downward, the other is moved upward: the down-going aileron increases the lift on its wing while the up-going aileron reduces the lift on its wing, producing a rolling (also called 'banking') moment about the aircraft's longitudinal axis (which extends from the nose to the tail of an airplane). Ailerons are usually situated near the wing tip, but may sometimes also be situated nearer the wing root. Modern airliners may also have a second pair of ailerons on their wings, and the terms 'outboard aileron' and 'inboard aileron' are used to describe these positions respectively.

A Yak-52 using ailerons to roll counter-clockwise during an aerobatic maneuver

An unwanted side effect of aileron operation is adverse yaw—a yawing moment in the opposite direction to the roll. Using the ailerons to roll an aircraft to the right produces a yawing motion to the left. As the aircraft rolls, adverse yaw is caused partly by the change in drag between the left and right wing. The rising wing generates increased lift, which causes increased induced drag. The descending wing generates reduced lift, which causes reduced induced drag. Profile drag caused by the deflected ailerons may add further to the difference, along with changes in the lift vectors as one rotates back while the other rotates forward.

A 1937 Waco VKS-7 cabin-class biplane with its pairs of quadruple ailerons linked by an external vertical connector to simplify its aileron control system. The ailerons on each side thus move either up or down together.

In a coordinated turn, adverse yaw is effectively compensated by the use of the rudder, which results in a sideforce on the vertical tail that opposes the adverse yaw by creating a favorable yawing moment. Another method of compensation is 'differential ailerons', which have been rigged such that the down-going aileron deflects less than the up-going one. In this case the opposing yaw moment is generated by a difference in profile drag between the left and right wingtips. *Frise ailerons* accentuate this profile drag imbalance by protruding beneath the wing of an upward-deflected aileron, most often by being hinged slightly behind the leading edge and near the bottom of the surface, with the lower section of the aileron surface's leading edge protruding slightly below the wing's undersurface when the aileron is deflected upwards, substantially increasing profile drag on that side. Ailerons may also be designed to use a combination of these methods.

With ailerons in the neutral position, the wing on the outside of the turn develops more lift than the opposite wing due to the variation in airspeed across the wing span, which tends to cause the aircraft to continue to roll. Once the desired angle of bank (degree of rotation about the longitudinal axis) has been obtained, the pilot uses opposite aileron to prevent the angle of bank from increasing due to this variation in lift across the wing span. This minor opposite use of the control must be maintained throughout the turn. The pilot also uses a slight amount of rudder in the same direction as the turn to counteract adverse yaw and to produce a "coordinated" turn wherein the fuselage is parallel to the flight path. A simple gauge on the instrument panel called the slip indicator, also known as "the ball", indicates when this coordination is achieved.

## Aileron Components

### Horns and Aerodynamic Counterbalances

Aileron horn visible extending from the wingtop on the top wing of a Fokker Dr. I

Particularly on larger or faster aircraft, control forces may be extremely heavy. Borrowing a discovery from boats that extending a control surface's area forward of the hinge lightens the forces needed first appeared on ailerons during World War I when ailerons were extended beyond the wingtip and provided with a horn ahead of the hinge. Known as overhung ailerons, possibly the best known examples are the Fokker Dr.I and Fokker D.VII. Later examples brought the counterbalance in line with the wing to improve control and reduce drag. This is seen less often now, due to the Frise type aileron which provides the same benefit.

### Trim Tabs

Trim tabs are small movable sections resembling scaled down ailerons located at or near the trailing edge of the aileron. On most propeller powered aircraft, the rotation of the propeller(s) induces a counteracting roll movement due to Newton's third law of motion, in that every action has an equal and opposite reaction. To relieve the pilot of having to provide continuous pressure on the stick in one direction (which causes fatigue) trim tabs are provided to adjust or *trim out* the pressure needed against any unwanted movement. The tab itself is deflected in relation to the aileron, causing the aileron to move in the opposite direction. Trim tabs come in two forms, adjustable and fixed. A fixed trim tab is manually bent to the required amount of deflection, while the adjustable trim tab can be controlled from within the cockpit so that different power settings or flight attitudes can be compensated for. Some large aircraft from the 1950s (including the Canadair Argus) used free floating control surfaces that the pilot controlled only through the deflection of trim tabs, in which case additional tabs were also provided to fine-tune the control to provide straight and level flight.

## Spades

Extra 300L aircraft, from below, showing light coloured quadrilateral spades at roughly midwing.

Spades are flat metal plates, usually attached to the aileron lower surface, ahead of the aileron hinge, by a lever arm. They reduce the force needed by the pilot to deflect the aileron and are often seen on aerobatic aircraft. As the aileron is deflected upward, the spade produces a downward aerodynamic force, which tends to rotate the whole assembly so as to further deflect the aileron upward. The size of the spade (and its lever arm) determines how much force the pilot needs to apply to deflect the aileron. A spade works in the same manner as a horn but is more efficient due to the longer moment arm.

## Mass Balance Weights

Mass balancing of the Frise-hinged ailerons on a Messerschmitt Bf 110 "zerstörer"

To prevent control surface flutter (aeroelastic flutter), the center of lift of the control surface should be behind the center of gravity of that surface. To achieve this, lead weights may be added to the front of the aileron. In some aircraft the aileron construction may be too heavy to allow this system to work without huge weight increases. In this case, the weight may be added to a lever arm to move the weight well out in front to the aileron body. These balance weights are tear drop shaped (to reduce drag), which make them appear quite different from spades, although both project forward and below the aileron. In addition to reducing flutter, mass balances also reduce the stick forces required to move the control surface in flight.

## Types of Ailerons

### Single Acting Ailerons

Used during aviation's pre-war "pioneer era" and into the early years of the First World War, these ailerons were each controlled by a single cable, which pulled the aileron up. When the aircraft was at rest, the ailerons hung vertically down. This type of aileron was used on the Farman III biplane 1909 and the Short 166. A "reverse" version of this, utilizing wing-warping, existed on the later version of the Santos-Dumont *Demoiselle*, which only warped the wingtips "downward". One of the disadvantages of this setup was a greater tendency to yaw than even with basic interconnected ailerons. During the 1930s a number of light aircraft used single acting controls but used springs to return the ailerons to their neutral positions when the stick was released.

### Wingtip Ailerons

The Blériot VIII with wingtip ailerons in 1908, deflected for a slight right bank.

Used on the first-ever airframe to have the combination of "joystick/rudder-bar" controls that directly led to the modern flight control system, the Blériot VIII in 1908, some designs of early aircraft used "wingtip" ailerons, where the entire wingtip was rotated to achieve roll control as a separate, pivoting roll-control surface—the AEA June Bug used a form of these, with both the experimental German Fokker V.1 of 1916 and the earlier versions of the Junkers J 7 all-duralumin metal demonstrator monoplane using them—the J 7 led directly to the Junkers D.I all-duralumin metal German fighter design of 1918, which had conventionally hinged ailerons. The main problem with this type of aileron is the dangerous tendency to stall if used aggressively, especially if the aircraft is already in danger of stalling, hence the use primarily on prototypes, and their replacement on production aircraft with more conventional ailerons.

### Frise Ailerons

Engineer Leslie George Frise (1897–1979) of the Bristol Aeroplane Company developed an aileron shape that is pivoted at about its 25 to 30% chord line and near its bottom surface , in order to decrease stick forces as aircraft became faster during the 1930s. When the aileron is deflected up (to make its wing go down), the leading edge of the aileron starts to protrude below the underside of the wing into the airflow beneath the wing. The moment of the leading edge in the airflow helps to move up the trailing edge, which decreases the stick force. The down moving aileron also adds energy to the boundary layer. The edge of the aileron directs air flow from the underside of the wing to the upper surface of the aileron, thus creating a lifting force added to the lift of the wing. This reduces the needed deflection of the aileron. Both the Canadian Fleet Model 2 biplane of 1930 and the 1938 popular US Piper J-3 Cub monoplane possessed Frise ailerons as designed and helped introduce them to a wide audience.

A claimed benefit of the Frise aileron is the ability to counteract adverse yaw. To do so, the leading edge of the aileron has to be sharp or bluntly rounded, which adds significant drag to the upturned aileron and helps counterbalance the yaw force created by the other aileron turned down. This can add some unpleasant, nonlinear effect and/or potentially dangerous aerodynamic vibration (flutter). Adverse yaw moment is basically countered by aircraft yaw stability and also by the use of differential aileron movement.

## Differential Ailerons

By careful design of the mechanical linkages, the up aileron can be made to deflect more than the down aileron (e.g., US patent 1,565,097). This helps reduce the likelihood of a wing tip stall when aileron deflections are made at high angles of attack. In addition, the consequent differential in drag reduces adverse yaw (as also discussed above). The idea is that the loss of lift associated with the up aileron carries no penalty while the increase in lift associated with the down aileron is minimized. The rolling couple on the aircraft is always the difference in lift between the two wings. A designer at de Havilland invented a simple and practical linkage and their de Havilland Tiger Moth classic British biplane became one of the best-known aircraft, and one of the earliest, to use differential ailerons.

## Roll Control without Ailerons

## Wing Warping

On the earliest Pioneer Era aircraft, such as the Wright Flyer and the later, 1909-origin Blériot XI and Etrich Taube, lateral control was effected by twisting the outboard portion of the wing so as to increase or decrease lift by changing the angle of attack. This had the disadvantages of stressing the structure, being heavy on the controls, and of risking stalling the side with the increased angle of attack during a maneuver. By 1916, most designers had abandoned wing warping in favor of ailerons. Researchers at NASA and elsewhere have been taking a second look at wing warping again, although under new names. The NASA version is the X-53 Active Aeroelastic Wing while the United States Air Force tested the Adaptive Compliant Wing. It is probably significant that none of the aircraft designed after those experiments implemented this type of control.

## Differential Spoilers

Spoilers are devices that when extended into the airflow over a wing, disrupt the airflow and reduce the amount of lift generated. Many modern aircraft designs, especially jet aircraft, use spoilers in lieu of, or to supplement ailerons, such as the F4 Phantom II and Northrop P-61 Black Widow, which had almost full width flaps (there were very small conventional ailerons at the wingtips as well).

## Roll induced by Rudder

All aircraft with dihedral have some form of yaw-roll coupling to promote stability. Common trainers like the Cesssna 152/172 series can be roll controlled with rudder alone. The rudder of the Boeing 737 has more roll authority over the aircraft than the ailerons at high angles of attack. This led to two notable incidents when the rudder jammed in the fully deflected position causing rollovers Boeing 737 rudder issues.

Some aircraft such as the Fokker Spin and model gliders lack any type of lateral control. Those aircraft use a higher amount of dihedral than conventional aircraft. Deflecting the rudder gives yaw and a lot of differential wing lift, giving a yaw induced roll moment. This type of control system is most commonly seen in the Flying Flea family of small aircraft and on simpler 2-function (pitch and yaw control) glider models or 3-function (pitch, yaw and throttle control) model powered aircraft, such as radio-controlled versions of "Old Timer" free-flight engine-powered model aircraft.

## Other Methods

- Weight-shift control is widely used in hang gliders, powered hang gliders, and ultralight aircraft.
- Flight with disabled controls has been successful in a small number of aviation incidents.
- Reaction control valves as used in the Harrier jump jet family of military aircraft.
- Top rudder: this device was fitted to the British Army Aeroplane No 1. It comprised an all-flying fin mounted above the upper wing and pivoted about a vertical axis. In operation it applied a side force approximately above the centre of pressure, causing the craft to roll. The design also had all-flying ailerons between the wing planes, but these were removed at the time it made the first official flight of a British aircraft and roll control during the flight was achieved solely by use of the top rudder.

## Elevator

An elevator is a primary flight control surface that controls movement about the lateral axis of an aircraft. This movement is referred to as "pitch". Most aircraft have two elevators, one of which is mounted on the trailing edge of each half of the horizontal stabilizer. When a manual or autopilot control input is made, the elevators move up or down as appropriate. In most installations, the elevators move symmetrically but, in some fly-by-wire controlled aircraft, they move differentially when required to meet the control input demands. Some aircraft types have provisions to "disconnect" the right and left elevators from one another in the event of a control surface jam while other types use different hydraulic systems to power the left and right elevator to ensure at least one surface is operational in the event of hydraulic system failure(s).

The elevator is used to control the position of the nose of the aircraft and the angle of attack of the wing. Changing the inclination of the wing to the local flight path changes the amount of lift which the wing generates. This, in turn, causes the aircraft to climb or dive. During take off the elevators are used to bring the nose of the aircraft up to begin the climb out. During a banked turn, elevator inputs can increase the lift and cause a tighter turn. That is why elevator performance is so important for fighter aircraft.

The elevators work by changing the effective shape of the airfoil of the horizontal stabilizer. Changing the angle of deflection at the rear of an airfoil changes the amount of lift generated by the foil. With greater downward deflection of the trailing edge, lift increases. With greater upward deflection of the trailing edge, lift decreases and can even become negative as shown on this slide. The lift force (F) is applied at center of pressure of the horizontal stabilzer which is some distance (L) from the aircraft center of gravity. This creates a torque

$$T = F * L$$

on the aircraft and the aircraft rotates about its center of gravity. The pilot can use this ability to make the airplane loop. Or, since many aircraft loop naturally, the deflection can be used to trim or balance the aircraft, thus preventing a loop. If the pilot reverses the elevator deflection to down, the aircraft pitches in the opposite direction.

## Elevator Control Effectiveness

The horizontal stabilizer usually creates a *downward* force which balances the nose down moment created by the wing lift force, which typically applies at a point (the wing center of lift) situated aft of the airplane's center of gravity. The effects of drag and changing the engine thrust may also result in pitch moments that need to be compensated with the horizontal stabilizer.

Both the horizontal stabilizer and the elevator contribute to pitch stability, but only the elevators provide pitch control. They do so by decreasing or increasing the downward force created by the stabilizer:

- An increased downward force, produced by *up* elevator, forces the tail down and the nose up. At constant speed, the wing's increased angle of attack causes a greater lift to be produced by the wing, accelerating the aircraft upwards. The drag and power demand also increase;
- A decreased downward force at the tail, produced by *down* elevator, causes the tail to rise and the nose to lower. At constant speed, the decrease in angle of attack reduces the lift, accelerating the aircraft downwards.

On many low-speed aircraft, a trim tab is present at the rear of the elevator, which the pilot can adjust to eliminate forces on the control column at the desired attitude and airspeed. Supersonic aircraft have all-moving tailplanes (stabilators), because shock waves generated on the horizontal stabilizer greatly reduced the effectiveness of hinged elevators. Delta winged aircraft combine ailerons and elevators –and their respective control inputs– into one control surface called an elevon.

## Elevators' Location

Elevators are usually part of the tail, at the rear of an aircraft. In some aircraft, pitch-control surfaces are in the front, ahead of the wing. In a two-surface aircraft this type of configuration is called a canard (the French word for duck) or a tandem wing. The Wright Brothers' early aircraft were of the canard type; Mignet Pou-du-Ciel and Rutan Quickie are of tandem type. Some early three surface aircraft had front elevators (Curtiss/AEA June Bug); modern three surface aircraft may have both front (canard) and rear elevators (Grumman X-29).

## Flaperon

Flaperons are a specialised type of aircraft flight control surface that combine aspects of both flaps and ailerons. Flaperons respond to roll commands, made by the pilot via the control wheel or stick, in the same manner as conventional ailerons. However, the flaperons can also be symmetrically lowered to function similarly to a dedicated set of flaps. Pilot controls for lowering or raising the flaperons are separate from those controlling the aileron function. A mixer is incorporated in the design to combine the two pilot inputs appropriately.

The use of flaperons instead of separate flaps and ailerons can result in an overall reduction in the weight of an aircraft. Flaperons are often mounted well clear of the trailing edge of the wing to ensure undisturbed airflow across the flight control surface whilst at low speed or at high angles of attack. This helps to reduce stall speed and improves low speed handling characteristics.

Flaperons were incorporated in a number of World War II era designs including the Junkers Ju87 Stuka dive bomber. Current utilisation is limited and is largely confined to light, experimental category (homebuilt) aircraft. Flaperons are also incorporated into the design of the V-22 Osprey tilt rotor aircraft.

## Operation

In addition to controlling the roll or bank of an aircraft, as do conventional ailerons, both flaperons can be lowered together to function similarly to a set of flaps.

On a plane with flaperons, the pilot still has the standard separate controls for ailerons and flaps, but the flap control also varies the flaperon's range of movement. A mechanical device called a "mixer" is used to combine the pilot's input into the flaperons. While the use of flaperons rather than ailerons and flaps might seem to be a simplification, some complexity remains through the intricacies of the mixer.

Some aircraft, such as the Denney Kitfox, suspend the flaperons below the wing (rather in the manner of slotted flaps) to provide undisturbed airflow at high angles of attack or low airspeeds. When the flaperon surface is hinged below the trailing edge of a wing, they are sometimes named "Junker Flaperons", from the *doppelflügel* (lit., "double wing") type of trailing edge surfaces used on a number of Junkers aircraft of the 1930s, such as the Junkers Ju 52 airliner, and Junkers Ju 87 *Stuka* iconic World War II dive bomber.

## Research

Current research seeks to coordinate the functions of aircraft flight control surfaces (ailerons, elevators, elevons, flaps, and flaperons) so as to reduce weight, cost, drag, inertia, and thereby achieve improved control response), reduced complexity, and reduced radar visibility for stealth purposes. Beneficiaries of such research might include drones (UAVs) and the latest fighter aircraft.

These research approaches include flexible wings and fluidics:

## Flexible Wings

In flexible wings, much or all of a wing surface can change shape in flight to deflect air flow. The X-53 Active Aeroelastic Wing is a NASA effort. The Adaptive Compliant Wing is a military and commercial effort. This may be seen as a return to the wing warping used and patented by the Wright brothers.

## Fluidics

In fluidics, forces in vehicles occur via circulation control, in which larger, more complex mechanical parts are replaced by smaller simpler fluidic systems (slots which emit air flows), where larger forces in fluids are diverted by smaller jets or flows of fluid intermittently, to change the direction of vehicles. In this use, fluidics promises lower mass, costs (up to 50% less), and very low inertia and response times, as well as simplicity.

## Air Brake or Speed Brake

Air brakes are defined as deployable devices which increase aerodynamic drag, thereby decelerating an aircraft. Their main objective is to decrease air speed in the flight phases of descent and approach, and to decrease landing distance. Military aircraft make extensive use of air brakes for tactical maneuvering. The use of air brakes when conducting steep approach is, at present, essential. Although the term air brake is sometimes used to refer to a specific type of vertically extending air brake, it also refers to any brake system that uses air flow to fulfill its duty. It is the latter, wider definition that is appealed to in this work. The range of major air-brake types in current usage is shown in figure.

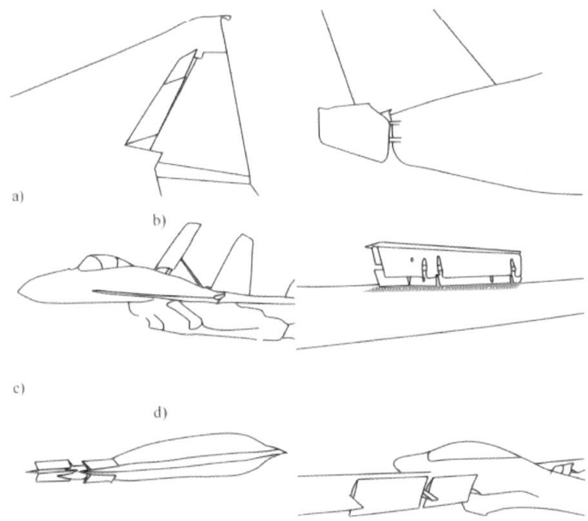


Major air-brake types in current usage. a) Wing spoiler of a CTA; b) Posterior/tail-mounted split air brake; c) Fuselage mounted dorsal air brake on fighter aircraft; d) Double-decker Schempp-Hirth air-brake; e) Deceleron; f) Turning flap.

The naming of air brakes is dependent on their position and is underpinned by an anatomical model, illustrated in Figure.

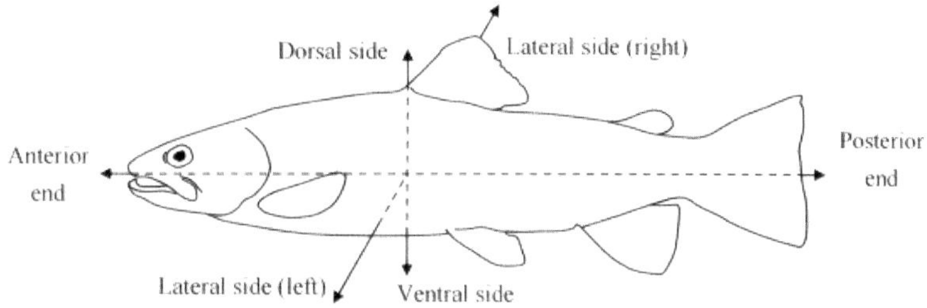


Anatomical model (trout) underpinning naming of air brakes.

The majority of air-brake types are wing-based. The most common are conventional Wing Spoilers (CS). These are used with nearly every contemporary CTA. Fowler Flaps are also used as air brakes with every contemporary CTA. Although their principal function is to increase the lift coefficient, they act as air brakes in their highest setting: their flap-extension range occupies half of their capability, generating lift proportional to induced drag. As the flaps extend beyond halfway, the inverse occurs, because profile drag at constant lift increases with profile camber and flap deflection X. Schempp-Hirth air-brakes, often found on gliders, are flat, sometimes perforated plates that extend perpendicular to the wing's surface. Some designs extend above and below the wing.

When investigated for their role in steep approaches, Mini Trailing Edge Devices (TEDs) have shown to carry an air-braking effect, in addition to lift increase. Examples of other integrated air-brake high-lift concepts include an integrated spoiler and flap system described in Ref.  and drooped spoilers with trailing-edge wedges .

Fuselage mounted air-brake systems have also been developed. The Embraer E-170 has a ventral air brake installed at the wing/fuselage fairing for use in steep approach. Dorsal air brakes are mainly used by fighter aircraft such as the Eurofighter Typhoon or MiG-29. While Posterior or tail-mounted split air brakes are used by the BAE 146-300, and the Fokker F-28 for steep approach.

An air brake is a panel conforming the shape of an aircraft that can be opened with hydraulic pressure in order to create drag, similar to spoilers which are on the edges of the aircraft wings and open in an upward position forcing the plane towards the ground. Air brake used in any circumstance when the aircraft needs to reduce its airspeed, however spoilers are only able to be opened when the airplane is approaching the runway and about to touch down. Lift dumpers, a type of air brake, is mounted on the top of a fuselage. When the panel is opened, it acts as a small spoiler, gently pushing the aircraft down. The air speed reducing devices found on the wings of aircraft, often called flaps have several different uses and designs. Most all flaps provide some amount of lift in addition to drag either for taking off or for keeping the nose up when landing. The slotted and fowler flaps are "both designed to physically increase the overall surface area of the wing, literally making the wing bigger". In slotted flaps, there are at least one open gap or slot in between the actual fixed wing of the aircraft and the adjustable flap. With these types, air passing underneath the wing of the aircraft is allowed to escape up through the slot due to the position of the flap tilting down. This directional flow of air and the extra surface area underneath the wing causes slight downforce pressure as well as reduced airspeed. Used more temporarily and for lower amounts of time than typical flaps, spoilers are similar to those seen on race cars. Spoilers are essentially an upside-down wing or lip causing downforce and drag. Spoilers on aircraft are only used when landing, not in midair, often with conjunction of slotted flaps. Together the flaps and spoiler disrupt the smooth flow of the wind and force it up and down across different surfaces protruding from or through the wing, causing a large amount of wind resistance and drag reducing overall air speed, acting as an air brake.

## Speedbrakes

Speedbrakes are high drag devices that are fitted to almost all high performance military aircraft as well as to some commercial aircraft types. In most cases, speedbrakes are fuselage mounted panels which, when selected by the pilot, extend into the airstream to produce drag. Dependant upon the aircraft type, the speed brake(s) may consist of a single panel or symetrically mounted pairs of panels. On the BA146, the speedbrakes are mounted on the tailcone. Speedbrakes may be used during the final approach to touchdown as well as after landing.

On many spoiler equiped aircraft, some of the spoiler panels have a flight spoiler function which is often referred to as "speedbrakes". In this application, the wing panels are symetrically extended by pilot selection. The maximum deflection of the panels while airborne is normally limited to an angle which is less than the deflection acheived in ground spoiler mode. Various aircraft have built in protections that will automatically command speedbrake retraction below a certain airspeed, with flaps selected beyond a given position or with thrust levers set above a specific angle. Wing spoilers should not be deployed during the final phase of the approach to landing as the induced loss of lift will result in a higher than normal stall speed and could result in a hard landing.

## Scientific Principles

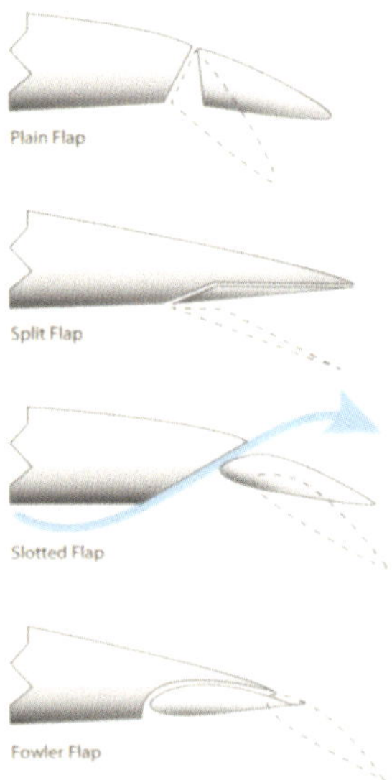


Examples of different styles of brake flap, with air flow around or through them

Principles such as plotting semi-empirical relations in order to estimate aircraft component parasite drag are necessary for developing and understanding the physical properties behind flight. Essentially, everything on the aircraft is taken into consideration for its size, surface area and amount of drag created. Everything from the location and size of the jet engines, to the shape of the fuselage, to wings and winglets and flaps are all noted, and equations are derived to develop an idea on the drag being created. Specifically, on the subject of air brakes and the intentional act of reducing air speed, slotted fowler flaps are examined, and a graph is constructed on the angle degree in which the flaps are tilted and the amount of lift and drag created by the distortion in the direction of air flow. Additionally, to reduce air speed, "whether in combat or landing, flat plates are used which are attached to the fuselage and shaped to its geometric contour when retracted". In the circumstance of air brakes used during subsonic flight, the equation for drag would be CDπ_brake=1.2-2.0 (with an average of 1.6).

## Technology

There are several styles of wing-based flaps and spoilers found on today's aircraft that greatly resemble some of the first air speed brakes invented. Slotted and fowler flaps are "both designed to physically increase the overall surface area of the wing, literally making the wing bigger". In slotted flaps, there at least one open gap or slot in between the actual fixed wind of the aircraft and the adjustable flap. With these types, air passing underneath the wing of the aircraft is allowed to escape up through the slot due to the position of the flap tilting down. This directional flow of air and the extra surface area underneath the wing causes slight downforce pressure as well as reduced airspeed. Spoilers are often used in larger aircraft and used more temporarily and for less amounts of time than typical flaps as they are designed to not only slow the craft down, but it creates downforce pushing the plane towards the ground. Spoilers are essentially an upside-down wing where the length on the top of the wing exceeds that of the bottom, resulting in the high air pressure on top of the spoiler to push down towards that of lower air pressure underneath. Lips are also used for the air running across the top of a wing to be forced further up, creating more drag and also pushing the plane closer to the ground. Spoilers on aircraft are only used when landing, not in midair, often with conjunction of slotted flaps. Together the flaps and spoiler disrupt the smooth flow of the wind and force it up and down across different surfaces protruding from or through the wing, causing a large amount of wind resistance and drag reducing overall air speed, acting as an air brake.

## Split Control Surfaces

The deceleron is an aileron that functions normally in flight but can split in half such that the top half goes up as the bottom half goes down to brake. This technique was first used on the F-89 Scorpion and has since been used by Northrop on several aircraft, including the B-2 Spirit.

Space Shuttle Discovery on landing, showing its rudder deployed in speed brake mode

The space shuttle used a similar system. The vertically-split rudder opened in "clamshell" fashion on landing to act as a speed brake, as shown in the photograph opposite.

## V-tail

Throughout the history of aviation there have been a number of very successful aircraft that employed V tails. They run the gamut from gliders to ultralights to midget racers to mainstream piston and jet aircraft~ and mainly fly just like their conventional-tail counterparts. In recent years, however, the Beech Bonanza received some negative press and was the subject of some FAA action based on a growing (but sparse) history of in-flight break-ups.

Aerodynamically, the V tail provides the same stabilizing forces in both the pitch and yaw axes that the conventional tail does. This is because the V tail has projected area in both directions. The "top view" of the tail represents the equivalent area of a flat horizontal tail, and the "side view" of the tail represents half the equivalent projected area of a vertical tail (this is because each side of the V tail provides an independent contribution to yaw stability.) V tails can be all-flying (stabilator style) or comprised of a fixed and moveable portion.

A V-Tail aircraft design incorporates two slanted tail surfaces instead of the horizontal and vertical fins of a conventional aircraft empennage. The two fixed tail surfaces act as both horizontal and vertical stabilisers and each has a moveable flight control surface referred to a ruddervator. These ruddervators perform the combined functions of both a rudder and an elevator.

When the pilot moves the control column forward or aft, the ruddervators move symmetrically in the same manner as a conventional elevator. Conversely, when the rudder pedals are displaced, the ruddervator surfaces move differentially to emulate the movement of a conventional rudder. when both the rudder pedals and the control column are moved simultaneously, an incorporated mixing mechanism moves each surface an appropriate amount.

The complexity of the required control system, the susceptibility of the design to Dutch roll and the minimal reduction in drag associated with a V-Tail design has limited its use. The FOUGA CM-175 Zephyr and the Beech BE35 Bonanza are two of the few aircraft types which incorporated a V-Tail design.

### Variants

The X-shaped tail surfaces of the experimental Lockheed XFV were essentially a V tail that extended both above and below the fuselage.

### Conventional

Despite its advantages, the V-tail has not become popular on aircraft design. The most popular conventionally V-tailed aircraft that was mass-produced is the Beechcraft Bonanza Model 35, often known as the *V-tail Bonanza* or simply *V-Tail*. Other examples include the F-117 Nighthawk stealth fighter and the Fouga Magister trainer. Some gliders, like PIK-16 Vasama, were designed with a V-tail, but the production Vasamas had cruciform tail.

### Inverted

The Blohm & Voss P 213 *Miniaturjäger* was one of the first aircraft having an inverted v-tail. Unmanned aerial vehicles such as the Amber, GNAT and the MQ-1 Predator would later feature this type of tail. The Ultraflight Lazair ultralights, of which over 2000 were produced, featured an inverted V-tail, which also carried the rear landing gear.

## Advantages

Ideally, with fewer surfaces than a conventional three-aerofoil tail or a T-tail, the V-tail is lighter, has less wetted surface area, so thus produces less induced and parasitic drag. However, NACA studies indicated that the V-tail surfaces must be larger than simple projection into the vertical and horizontal planes would suggest, such that total wetted area is roughly constant; reduction of intersection surfaces from three to two does, however, produce a net reduction in drag through elimination of some interference drag.

In modern day, light jet general aviation aircraft such as the Cirrus Vision, Eclipse 400 or the unmanned aerial drone Global Hawk often have the power plant placed outside the aircraft to protect the passengers and make certification easier. In such cases V-tails are used to avoid placing the vertical stabilizer in the exhaust of the engine, which would disrupt the flow of the exhaust, reducing thrust and increasing wear on the stabilizer, possibly leading to damage over time.

## Disadvantages

In the mid-1980s, the Federal Aviation Administration grounded the Beechcraft Bonanza due to safety concerns. While the Bonanza met the initial certification requirements, it had a history of fatal mid-air breakups during extreme stress, at a rate exceeding the accepted norm. The type was deemed airworthy and restrictions removed after Beechcraft issued a structural modification as an Airworthiness Directive.

V-tailed aircraft require longer rear fuselage than aircraft with conventional empennage to prevent yawing. This tendency, called "snaking", was apparent on Fouga Magister (which has a relatively short fuselage) on taking off and landing.

## Ruddervators

A top-down view of the Northrop YF-23 *Gray Ghost* prototype fighter jet, showing its distinctive wide V-tail and Ruddervators.

Ruddervators are the control surfaces on an airplane with a V-tail configuration. They are located at the trailing edge of each of the two airfoils making up the tail of the plane. The first use of ruddervators may have been on the Coandă-1910's X-tail, although there is no proof that the aircraft ever flew. The later Coandă-1911 flew with ruddervators on its X-tail. Later Polish engineer Jerzy Rudlicki designed the first practical ruddervators in 1930, tested on a modified Hanriot H-28 trainer in 1931.

The name is a portmanteau of the words rudder and elevator. In a conventional aircraft tail configuration, the rudder provides yaw (horizontal) control and the elevator provides pitch (vertical) control.

Ruddervators provide the same control effect as conventional control surfaces, but through a more complex control system that actuates the control surfaces in unison. Yaw moving the nose to the left is produced on an upright V tail by moving the pedals left which deflects the left-hand ruddervator down and left and the right-hand ruddervator up and left. The opposite produces yaw to the right. Pitch nose up is produced by moving the control column or stick back which deflects the left-hand ruddervator up and right and the right-hand ruddervator up and left. Pitch nose down is produced by moving the control column or stick forward which induces the opposite ruddervator movements.

## Trim Tab

Trim tabs are small, secondary flight control surfaces that are attached to the trailing edge of a larger, primary control surface such as an elevator or rudder. The trim tab is used to "trim", or counteract the aerodynamic force on, the control surface and thus stabilise the aircraft in the axis of rotation associated with the primary control. Trim tabs can be either flight adjustable or fixed.

By adjusting the angle of the trim tab relative to the larger control surface, the neutral position of the control surface is changed. This occurs because the distance from the control surface hinge at which the trim tab is mounted provides sufficient moment or force to move the larger surface.

Fixed trim tabs are most often encountered on the rudder or aileron of a light, single engine aircraft. A fixed tab is ground adjusted (bent) to the desired position to overcome any unwanted roll or yaw tendencies of the aircraft. Adjustable trim is virtually always installed to adjust control loading associated with the pitch axis and may also be used to trim for yaw or roll. The position of an adjustable trim tab can be changed in flight, using cockpit controls, to counter any variation in control loading resulting from changes in airspeed, centre of gravity, aircraft configuration or due to an engine failure. Trim tabs can be manually or electrically controlled (or both), depending upon the aircraft installation.

## Uses in Aircraft

All aircraft must have a system for ensuring trim in the pitch axis, though methods other than trim tabs may be used. Alternatives include:

- A spring included in the control system that can be adjusted by the pilot.
- In the case of the elevator, an all-moving horizontal stabilizer, called a stabilator, the position of which can be adjusted in flight by a servo tab or an anti-servo tab.
- On some aircraft (e.g. Concorde, McDonnell Douglas MD-11), fuel may be shifted to tanks

in the tail during cruise to reposition the center of gravity in order to reduce trim drag. Maintaining the cg near the most aft limit for cruise improves cruise efficiency.

Elevator trim frees the pilot from exerting constant pressure on the pitch controls. Instead, the pilot adjusts a longitudinal trim control (often in the form of a wheel) to cancel out control forces for a given airspeed / weight distribution. Typically, when this trim control (wheel or lever) is rotated or moved forward, the nose is held down; conversely, if the trim control is moved back, the tail becomes "heavy". Many newer aircraft, especially jet aircraft, have electric trim controls.

Many airplanes also have rudder and/or aileron trim systems. On some of these, the rudder trim tab is rigid but adjustable on the ground by bending: it is angled slightly to the left (when viewed from behind) to lessen the need for the pilot to push the rudder pedal constantly to overcome the left-turning tendencies of many prop-driven aircraft. Some aircraft have hinged rudder trim tabs that the pilot can adjust in flight.

When a servo tab is employed, it is moved into the slipstream opposite to the control surface's desired deflection. For example, in order to trim an elevator to hold the nose down, the elevator's trim tab will actually rise *up* into the slipstream. The increased pressure on top of the trim tab surface caused by raising it will then deflect the *entire* elevator slab down slightly, causing the tail to rise and the aircraft's nose to move down. In the case of an aircraft where deployment of high-lift devices (flaps) would significantly alter the longitudinal trim, a supplementary trim tab is arranged to simultaneously deploy with the flaps so that pitch attitude is not markedly changed.

The use of trim tabs significantly reduces pilots' workload during continuous maneuvers (e.g. sustained climb to altitude after takeoff or descent prior to landing), allowing them to focus their attention on other tasks such as traffic avoidance or communication with air traffic control.

Both *elevator trim* and *pitch trim* affect the small trimming part of the elevator on jet airliners. The former is supposed to be set in a certain position for a longer time, while the pitch trim (controlled with the landing pilot's thumb on the yoke or joystick, and is thereby easy to maneuver) is used all the time after the flying pilot has disabled the autopilot, especially after each time the flaps are lowered or at every change in the airspeed, at the descent, approach and final. Elevator trim is most used for controlling the attitude at cruising by the autopilot.

Beyond reducing pilot workload, proper trim also increases fuel efficiency by reducing drag. For example, propeller aircraft have a tendency to yaw when operating at high power, for instance when climbing; this increases parasitic drag because the craft is not flying straight into the apparent wind. In such circumstances, the use of an adjustable rudder trim tab can reduce yaw.

On military aircraft during wartime, trim tabs often served as unintentional backup control systems for aircraft with damaged controls. Since trim tabs are usually controlled by their own dedicated system of control cables, rods, and/or hydraulic lines, aircraft that had suffered loss of primary controls could often be flown home "on the trim tabs", or by using trim adjustment as a replacement for the non-working primary controls. Such control is effective, if slower and more limited than primary controls, but it does allow the aircraft to be controlled and directed. In other cases, such as an engine being shot out, or damage causing asymmetric drag, trim tabs were invaluable for allowing the pilot to fly the aircraft straight without having to apply a constant force on the stick or rudder to keep the aircraft flying straight. Trim tabs were also important for aircraft such as bombers, who often underwent rapid changes in center-of-gravity when the bombload was dropped, requiring a hand ready on the trim-adjusting wheel to counteract the tendency of the aircraft to pitch up or down. Undertaking high-speed dives or deploying flaps also generally necessitated pitch trim adjustment, as aircraft of the era had different pitch tendencies at different airspeeds, and flaps could change the center of lift. Even burning fuel off could require periodic trim adjustment during a long flight, as it was difficult to ensure that all fuel tanks were equally near the center of gravity (careful selection of various fuel tanks to ensure even consumption from all tanks helped negate this as well). An extreme example would be later P-51 Mustangs, which were given a large long-range fuel tank behind the cockpit to allow long range missions to Berlin. Since the aircraft was not designed to carry this weight so far back in the fuselage, the trim had to be adjusted far forward during the first leg of the flight, and slowly adjusted back as the fuel was used, to avoid causing excess drag, and thus excessive fuel consumption, due to a nose-high attitude on the many-hour flight to Berlin.

## Types of Trim Tabs

There are four main types of trim tabs:

- Trim tabs
- Balance tabs
- Antiservo tabs
- Ground adjustable tabs

So now that you know the four types, let's look at how each of them work.

## Trim Tab

The plain old trim tab is one of the most common types of tabs used in small single-engine airplanes. A trim tab is attached to the trailing edge of an elevator, and it's operated by moving a small control wheel in the cockpit.

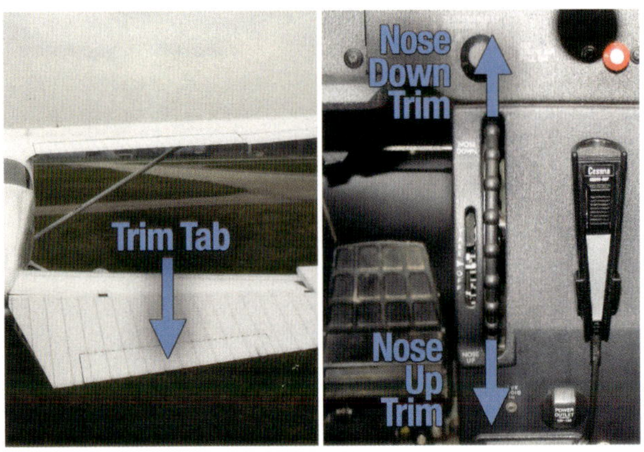


Night Flier

The operation is pretty simple: roll the wheel in the nose up position, and the tab moves down. Roll the wheel in the nose down direction, and the tab moves up.

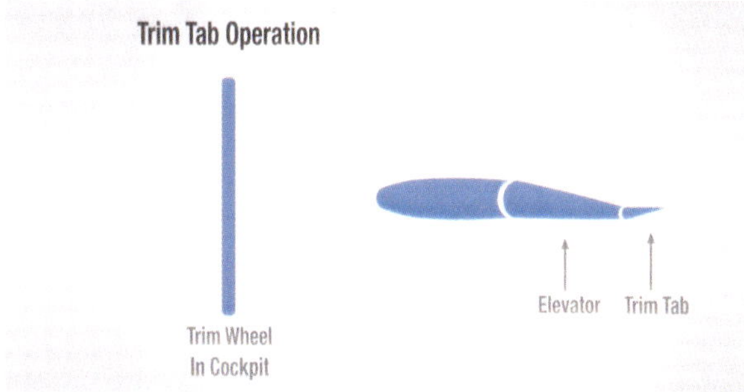


When you move the trim tab up or down, it sticks out into the free air stream, and deflects the elevator in the opposite direction. So even though it may seem 'backwards' to move the tab down to make the nose of the plane go up, seeing the tab in action helps it all make sense.

## Balance Tab

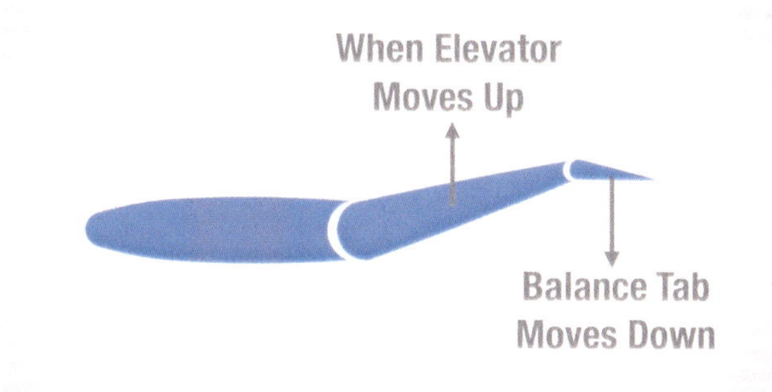


Some aircraft have very heavy control loads, especially at high speeds. That's where balance tabs come in handy. Balance tabs look like trim tabs, but they have one major difference: balance tabs are attached to the control surface linkage, so when the control surface is moved in one direction, the balance tab moves in the opposite direction.

By moving the balance tab in the opposite direction, the control load on your yoke is significantly reduced, making your airplane easier to fly.

## Antiservo Tab

Antiservo tabs are similar to balance tabs, but they move in the opposite direction. For example, when your elevator or stabilator moves up, the antiservo tab moves in the same direction.

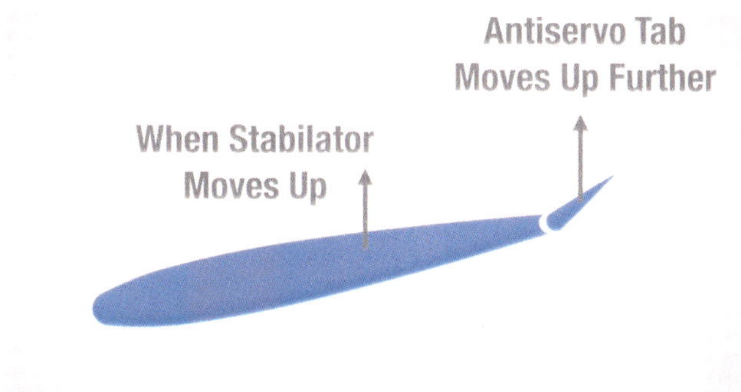

So why would you want your tab to move in the same direction as your elevator/stabilator? In small aircraft, it increases the control feel, and helps prevent you from over-controlling your aircraft's pitch. One of the most popular examples of the antiservo tab is on the Piper Cherokee. Without it, the plane would be much easier to pitch up and down, but it would also be easy to over-control, and possibly overstress the airframe..

Adrian Milliner

## Ground Adjustable Tab

The fourth and final tab is the ground adjustable tab. If you've flown a training airplane, there's a good chance it had one of these on the rudder. Ground adjustable tabs are just that: only adjustable on the ground. So how do you adjust it? By bending it left or right, preferably between two solid surfaces, like blocks of wood. But before you run out and start adjusting the tab yourself, it's a good idea to see what your POH or mechanic recommends first.

Markus Wichmann

The ground tab is used to keep your plane flying coordinated in level flight. And since it's a trial and error system, it might take a few flights of adjusting, flying, and more adjusting. But once you've got it set, you're good to go.

## Flap

Flaps are a high lift device consisting of a hinged panel or panels mounted on the trailing edge of the wing. When extended, they increase the camber and, in most cases, the chord and surface area of the wing resulting in an increase of both lift and drag and a reduction of the stall speed. These factors result in an improvement in takeoff and landing performance.

There are many different flap designs and configurations in use. Large aircraft sometimes incorporate more than one type, utilising different flap designs on the inboard and outboard sections of the wing. The following are descriptions of some of the more common flap designs:

- Plain Flap - The rear portion of the wing aerofoil rotates downwards on a simple hinge arrangement mounted at the front of the flap.
- Split Flap - The rear portion of the lower surface of the wing aerofoil hinges downwards from the leading edge of the flap, while the upper surface remains immobile.
- Slotted Flap - Similar to a Plain Flap but incorporates a gap between the flap and the wing to force high pressure air from below the wing over the upper surface of the flap. This helps reduce boundary layer separation and allows the airflow over the flap to remain laminar.
- Fowler Flap - A split flap that slides rearwards level for a distance prior to hinging downwards. It thereby first increases chord (and wing surface area) and then increases camber. This produces a flap which can optimise both takeoff (partial extension for optimal lift) and landing (full extension for optimal lift and drag) performance. This type of flap or one of its variations is found on most large aircraft.
- Double Slotted Fowler Flap - This design improves the performance of the Fowler flap by incorporating the boundary layer energising features of the slotted flap.

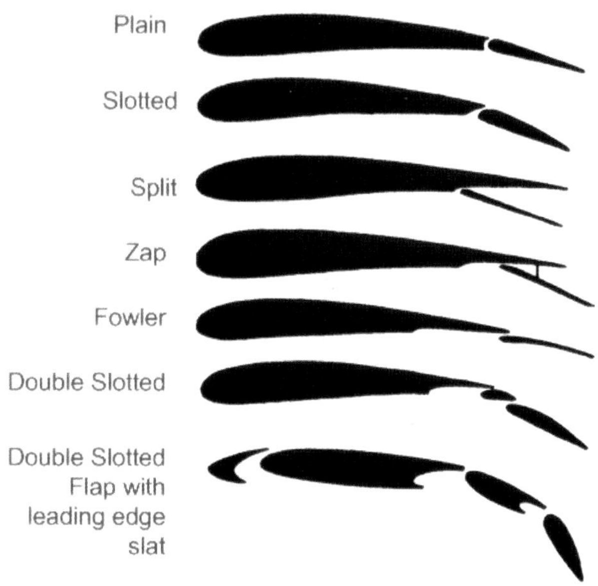


Flap Types

## Principles of Operation

The general airplane lift equation demonstrates these relationships:

$$L = \frac{1}{2}\rho V^2 S C_L$$

where:

- *L* is the amount of *Lift* produced,
- $\rho$ is the air density,
- *V* is the true airspeed of the airplane or the *Velocity* of the airplane, relative to the air,
- S is the area of the wing,
- $C_L$ is the *lift coefficient*, which is determined by the shape of the airfoil used and the angle at which the wing meets the air (or angle of attack).

Here, it can be seen that increasing the area (S) and lift coefficient ( $C_L$ ) allow a similar amount of lift to be generated at a lower airspeed (V).

Extending the flaps also increases the drag coefficient of the aircraft. Therefore, for any given weight and airspeed, flaps increase the drag force. Flaps increase the drag coefficient of an aircraft due to higher induced drag caused by the distorted spanwise lift distribution on the wing with flaps extended. Some flaps increase the wing area and, for any given speed, this also increases the parasitic drag component of total drag.

The three orange pods are fairings streamlining the flap track mechanisms. The flaps (two on each side, on the Airbus A319) lie directly above these.

## Flaps during Takeoff

Depending on the aircraft type, flaps may be partially extended for takeoff. When used during takeoff, flaps trade runway distance for climb rate: using flaps reduces ground roll but also reduces the climb rate. The amount of flap used on takeoff is specific to each type of aircraft, and the manufacturer will suggest limits and may indicate the reduction in climb rate to be expected. The *Cessna 172S Pilot Operating Handbook* generally recommends 10° of flaps on takeoff, especially when the ground is rough or soft.

## Flaps during Landing

Flaps may be fully extended for landing to give the aircraft a lower stall speed so the approach to landing can be flown more slowly, which also allows the aircraft to land in a shorter distance. The higher lift and drag associated with fully extended flaps allows a steeper and slower approach to the landing site, but imposes handling difficulties in aircraft with very low wing loading (i.e. having little weight and a large wing area). Winds across the line of flight, known as *crosswinds*, cause the windward side of the aircraft to generate more lift and drag, causing the aircraft to roll, yaw and pitch off its intended flight path, and as a result many light aircraft land with reduced flap settings in crosswinds. Furthermore, once the aircraft is on the ground, the flaps may decrease the effectiveness of the brakes since the wing is still generating lift and preventing the entire weight of the aircraft from resting on the tires, thus increasing stopping distance, particularly in wet or icy conditions. Usually, the pilot will raise the flaps as soon as possible to prevent this from occurring.

Flaps during ground roll after landing,with spoilers up, increasing drag.

North American T-6 trainer showing its split flaps

## Maneuvering Flaps

Some gliders not only use flaps when landing, but also in flight to optimize the camber of the wing for the chosen speed. While thermalling, flaps may be partially extended to reduce the stall speed so that the glider can be flown more slowly and thereby reduce the rate of sink, which lets the glider use the rising air of the thermal more efficiently, and to turn in a smaller circle to make best use of the core of the thermal. At higher speeds a negative flap setting is used to reduce the nose-down pitching moment. This reduces the balancing load required on the horizontal stabilizer, which in turn reduces the trim drag associated with keeping the glider in longitudinal trim. Negative flap may also be used during the initial stage of an aerotow launch and at the end of the landing run in order to maintain better control by the ailerons.

Like gliders, some fighters such as the Nakajima Ki-43 also use special flaps to improve maneuverability during air combat, allowing the fighter to create more lift at a given speed, allowing for much tighter turns. The flaps used for this must be designed specifically to handle the greater stresses and most flaps have a maximum speed at which they can be deployed. Control line model aircraft built for precision aerobatics competition usually have a type of maneuvering flap system that moves them in an opposing direction to the elevators, to assist in tightening the radius of a maneuver.

## Flap Tracks

Extending flaps often run on guide tracks. Where these run outside the wing structure they may be faired in to streamline them and protect them from damage. Some track fairings are designed to act as anti-shock bodies, which reduce drag caused by local sonic shock waves where the airflow becomes transonic at high speeds.

## Thrust Gates

Thrust gates, or gaps, in the trailing edge flaps may be required to minimise interference between the engine flow and deployed flaps. In the absence of an inboard aileron, which provides a gap in many flap installations, a modified flap section may be needed. The thrust gate on the Boeing 757 was provided by a single-slotted flap in between the inboard and outboard double-slotted flaps. The A320, A330, A340 and A380 have no inboard aileron. No thrust gate is required in the continuous, single-slotted flap. Interference in the go-around case while the flaps are still fully deployed can cause increased drag which must not compromise the climb gradient.

## Types of Flap

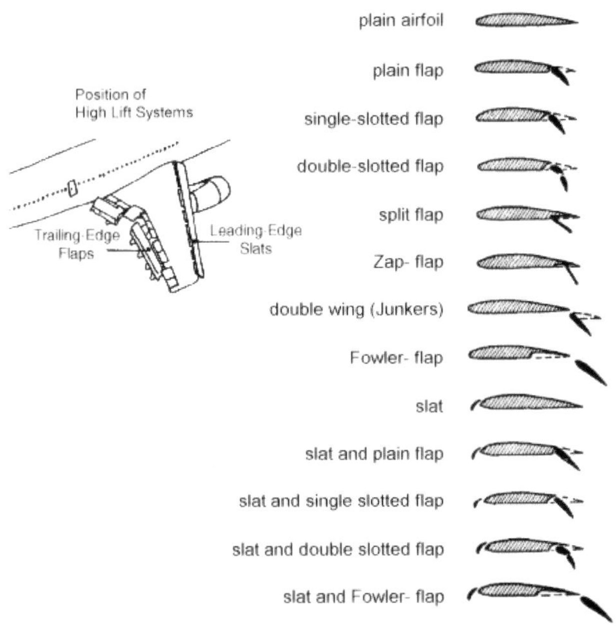


Flaps and high lift devices. Gurney flap exaggerated for clarity. Blown flap skipped as it is modified from any other type. Pale lines indicate line of movement, and green indicates flap setting used during dive.

## Plain Flap

The rear portion of airfoil rotates downwards on a simple hinge mounted at the front of the flap. The Royal Aircraft Factory and National Physical Laboratory in the United Kingdom tested flaps in 1913 and 1914, but these were never installed in an actual aircraft. In 1916, the Fairey Aviation Company made a number of improvements to a Sopwith Baby they were rebuilding, including their Patent Camber Changing Gear, making the Fairey Hamble Baby as they renamed it, the first aircraft to fly with flaps. These were full span plain flaps which incorporated ailerons, making it also the first instance of flaperons. Fairey were not alone however, as Breguet soon incorporated automatic flaps into the lower wing of their Breguet 14 reconnaissance/bomber in 1917. Due to the greater efficiency of other flap types, the plain flap is normally only used where simplicity is required.

## Split Flap

The rear portion of the lower surface of the airfoil hinges downwards from the leading edge of the flap, while the upper surface stays immobile. This can cause large changes in longitudinal trim, pitching the nose either down or up. At full deflection, a split flaps acts much like a spoiler, adding significantly to drag coefficient. It also adds a little to lift coefficient. It was invented by Orville Wright and James M. H. Jacobs in 1920, but only became common in the 1930s and was then quickly superseded. The Douglas DC-1 (progenitor to the DC-3 and C-47) was one of the first of many aircraft types to use split flaps.

## Slotted Flap

A gap between the flap and the wing forces high pressure air from below the wing over the flap helping the airflow remain attached to the flap, increasing lift compared to a split flap. Additionally, lift across the entire chord of the primary airfoil is greatly increased as the velocity of air leaving its trailing edge is raised, from the typical non-flap 80% of freestream, to that of the higher-speed, lower-pressure air flowing around the leading edge of the slotted flap. Any flap that allows air to pass between the wing and the flap is considered a slotted flap. The slotted flap was a result of research at Handley-Page, a variant of the slot that dates from the 1920s, but wasn't widely used until much later. Some flaps use multiple slots to further boost the effect.

## Fowler Flap

A split flap that slides backwards, before hinging downward, thereby increasing first chord, then camber. The flap may form part of the upper surface of the wing, like a plain flap, or it may not, like a split flap, but it must slide rearward before lowering. As a defining feature - distinguishing it from the Gouge Flap - it always provides a slot effect. Invented by Harlan D. Fowler in 1924, and tested by Fred Weick at NACA in 1932. They were first used on the Martin 146 prototype in 1935, and in production on the 1937 Lockheed Electra, and are still in widespread use on modern aircraft, often with multiple slots.

## Junkers Flap

A slotted plain flap where the flap is fixed below the trailing edge of the wing, rotating about its forward edge, and usually forming the "inboard" hinged section (closer to the root) of the Junkers

*Doppelflügel*, or "double-wing" style of wing trailing edge control surfaces (including the outboard-mounted ailerons), which hung just below and behind the wing's fixed trailing edge. When not in use, it has more drag than other types, but is more effective at creating additional lift than a plain or split flap, while retaining their mechanical simplicity. Invented by Otto Mader at Junkers in the late 1920s, they were most often seen on the Junkers Ju 52 and the Junkers Ju 87 *Stuka* dive bomber, though the same wing control surface can be also be found on many modern ultralights, like the Denney Kitfox.

### Gouge Flap

A type of split flap that slides backward along curved tracks that force the trailing edge downward, increasing chord and camber without affecting trim or requiring any additional mechanisms. It was invented by Arthur Gouge for Short Brothers in 1936 and used on the Short Empire and Sunderland flying boats, which used the very thick Shorts A.D.5 airfoil. Short Brothers may have been the only company to use this type.

### Fairey-youngman Flap

Drops down (becoming a Junkers Flap) before sliding aft and then rotating up or down. Fairey was one of the few exponents of this design, which was used on the Fairey Firefly and Fairey Barracuda. When in the extended position, it could be angled up (to a negative angle of incidence) so that the aircraft could be dived vertically without needing excessive trim changes.

### Zap Flap

Commonly, but incorrectly, called the Zapp Flap, it was Invented by Edward F. Zaparka while he was with Berliner/Joyce and tested on a General Aircraft Corporation Aristocrat in 1932 and on other types periodically thereafter, but it saw little use on production aircraft other than on the Northrop P-61 Black Widow. The leading edge of the flap is mounted on a track, while a point at mid chord on the flap is connected via an arm to a pivot just above the track. When the flap's leading edge moves aft along the track, the triangle formed by the track, the shaft and the surface of the flap (fixed at the pivot) gets narrower and deeper, forcing the flap down.

### Krueger Flap

A hinged flap which folds out from under the wing's leading edge while not forming a part of the leading edge of the wing when retracted. This increases the camber and thickness of the wing, which in turn increases lift and drag. This is not the same as a leading edge droop flap, as that is formed from the entire leading edge. Invented by Werner Krüger in 1943 and evaluated in Goettingen, Krueger flaps are found on many modern swept wing airliners.

### Gurney Flap

A small fixed perpendicular tab of between 1 and 2% of the wing chord, mounted on the high pressure side of the trailing edge of an airfoil. It was named for racing car driver Dan Gurney who rediscovered it in 1971, and has since been used on some helicopters such as the Sikorsky S-76B to correct control problems without having to resort to a major redesign. It boosts the efficiency of even basic theoretical airfoils (made up of a triangle and a circle overlapped) to the equivalent of a conventional airfoil. The principle was discovered in the 1930s, but was rarely used and was then forgotten. Late marks of the Supermarine Spitfire used a bead on the trailing edge of the elevators, which functioned in a similar manner.

## Leading Edge Flap

The entire leading edge of the wing rotates downward, effectively increasing camber and also slightly reducing chord. Most commonly found on fighters with very thin wings unsuited to other leading edge high lift devices.

## Blown Flap

A type of Boundary Layer Control System, blown flaps pass engine-generated air or exhaust over the flaps to increase lift beyond that attainable with mechanical flaps. Types include the original (internally blown flap) which blows compressed air from the engine over the top of the flap, the externally blown flap, which blows engine exhaust over the upper and lower surfaces of the flap, and upper surface blowing which blows engine exhaust over the top of the wing and flap. While testing was done in Britain and Germany before the Second World War, and flight trials started, the first production aircraft with blown flaps wasn't until the 1957 Lockheed T2V SeaStar. Upper Surface Blowing was used on the Boeing YC-14 in 1976.

## Flexible Flap

Also known as the FlexFoil. A modern interpretation of wing warping, internal mechanical actuators bend a lattice that changes the airfoil shape. It may have a flexible gap seal at the transition between fixed and flexible airfoils.

## Flaperon

A type of aircraft control surface that combines the functions of both flaps and ailerons.

## Continuous Trailing-edge Flap

As of 2014, U.S. Army Research Laboratory (ARL) researchers at NASA's Langley Research Center developed an active-flap design for helicopter rotor blades. The Continuous Trailing-Edge Flap (CTEF) uses components to change blade camber during flight, eliminating mechanical hinges in order to improve system reliability. Prototypes were constructed for wind-tunnel testing.

A team from ARL completed a live-fire test of a rotor blade with individual blade control technology in January 2016. The live fire experiments explored the ballistic vulnerability of blade control technologies.Researchers fired three shots representative of typical ground fire on a 7-foot span, 10-inch chord rotor blade section with a 4-foot long CTEF at ARL's Airbase Experimental Facility.

## Spoilers

Spoilers are panels mounted on the upper surface of the wing that, when extended, both increase drag and decrease lift by disrupting the airflow over the wing. Dependent upon the aircraft type, spoilers can serve as many as three distinct primary functions:

- Ground spoilers
- Roll spoilers
- (Flight) spoilers or Speedbrakes

Some aircraft such as the AIRBUS A-320 and the EMBRAER ERJ 190-100 have all three spoiler functions whereas the BA146 only incorporates the ground spoiler function. Certain aircraft designs also utilize the spoiler panels for secondary functions such as turbulence damping.

Spoilers are small, hinged plates on the top portion of wings. Spoilers can be used to slow an aircraft, or to make an aircraft descend, if they are deployed on both wings. Spoilers can also be used to generate a rolling motion for an aircraft, if they are deployed on only one wing.

### Spoilers Deployed on both Wings

When the pilot activates the spoilers, the plates flip up into the air stream. The flow over the wing is disturbed by the spoiler, the drag of the wing is increased, and the lift is decreased. Spoilers can be used to "dump" lift and make the airplane descend; or they can be used to slow the airplane down as it prepares to land. When the airplane lands on the runway, the pilot usually brings up the spoilers to kill the lift, keep the plane on the ground, and make the brakes work more efficiently. The friction force between the tires and the runway depends on the "normal" force, which is the weight minus the lift. The lower the lift, the better the brakes work. The additional drag of the spoilers also slows the plane down.

### Spoiler Deployed on only One Wing

A single spoiler is used to bank the aircraft; to cause one wing tip to move up and the other wing tip to move down. The banking creates an unbalanced side force component of the large wing lift force which causes the aircraft's flight path to curve. (Airplanes turn because of banking, not because of the force generated by the rudder).

### Spoiler Controls

Spoiler controls can be used for roll control (outboard or mid-span spoilers) or descent control (inboard spoilers).

Some aircraft use spoilers in combination with or in lieu of ailerons for roll control, primarily to reduce adverse yaw when rudder input is limited by higher speeds. For such spoilers the term spoileron has been coined. In the case of a spoileron, in order for it to be used as a control surface, it is raised on one wing only, thus decreasing lift and increasing drag, causing roll and yaw. Spoilerons also avoid the problem of control reversal that affects ailerons.

Almost all modern jet airliners are fitted with inboard lift spoilers which are used together during descent to increase the rate of descent and control speed. Some aircraft use lift spoilers on landing approach to control descent without changing the aircraft's attitude.

One jet airliner not fitted with lift spoilers was the Douglas DC-8 which used reverse thrust in flight on the two inboard engines to control descent speed (however the aircraft was fitted with lift dumpers). The Lockheed Tristar was fitted with a system called Direct Lift Control using the spoilers on landing approach to control descent.

Airbus aircraft with fly-by-wire control utilise wide-span spoilers for descent control, spoilerons, gust alleviation, and lift dumpers. Especially on landing approach, the full width of spoilers can be seen controlling the aircraft's descent rate and bank.

### Lift Dumpers

Lift dumpers are a special type of spoiler extending along much of the wing's length and designed to dump as much lift as possible on landing. Lift dumpers have only two positions, deployed and retracted. Lift dumpers have three main functions: putting most of the weight of the aircraft on the wheels for maximum braking effect, increasing form drag, and preventing aircraft 'bounce' on landing.

Lift dumpers are almost always deployed automatically on touch down. The flight deck control has three positions: off, automatic ('armed'), and manual (rarely used). On landing approach 'automatic' is selected and, at the moment of touchdown, lift dumpers are deployed in a fraction of a second, with flight control spoilers also being raised automatically as additional lift dumpers.

Virtually all modern jet aircraft are fitted with lift dumpers. The British Aerospace 146 is fitted with particularly wide span spoilers to generate additional drag and make reverse thrust unnecessary.

A number of accidents have been caused either by inadvertently deploying lift dumpers on landing approach, or forgetting to set them to 'automatic'.

## Aircraft Engine Controls

Many aircraft engines are geared to slow the engine RPMs down such that the propellers operate at an efficient RPM. Turboprops for example.

Most internal combustion propeller driven aircraft fly at specific RPMs for the different stages of flight. For example, take-offs and departures are usually at the designed maximum RPM, while enroute the engine is usually operated at a slower and more efficient RPM. You can think of the RPMs as acting kind of like gears in a car.

Interestingly, an aircraft engine's RPM might not be changed while enroute, yet the engine power (e.g. speed) CAN be changed easily.

There are three controls that are used to adjust an internal combustion engine's power. The throttle is very similar to a car's gas pedal in terms of adding or subtracting power. But an airplane is different in that the engine*typically* runs at the same speed even as the throttle changes the engine's power output.

How does that work? The propeller knob changes the engine's RPMs. For take-off it is set full forward to ensure maximum power when the throttle is also pushed full forward. So you see, a given RPM acts like a particular gear in a car and the throttle acts like the gas pedal.

Once at cruise altitude cruise speed is set by reducing the throttle to the appropriate cruise power setting (measured in inches of mercury), and then the propeller RPMs are also reduced to the cruise setting. Finally, the mixture knob is used to lean the fuel/air ratio to the correct stoichiometric ratio for the altitude. The reason for the mixture control that you won't find in a car is that aircraft operate at a range of altitude far beyond what most cars do and as such the mixture needs to be able to be manually adjusted.

## Basic Controls and Indicators

- Master Switch - Most often actually two separate switches, the Battery Master and the Alternator Master. The Battery Master activates a relay (sometimes called the battery contactor) which connects the battery to the aircraft's main electrical bus. The alternator master activates the alternator by applying power to the alternator field circuit. These two switches provide electrical power to all the systems in the aircraft.
- Throttle - Sets the desired power level. The throttle controls the mass flow-rate of air (in fuel-injected engines) or air/fuel mixture (in carburetted engines) delivered to the cylinders.
- Propeller Control - Adjusts the Constant Speed Unit, which in turn adjusts the propeller pitch and regulates the engine load as necessary to maintain the set R.P.M.
- Mixture Control - Sets the amount of fuel added to the intake airflow. At higher altitudes, the air pressure (and therefore the oxygen level) declines so the fuel volume must also be reduced to give the correct air/fuel mixture. This process is known as "leaning".
- Ignition Switch - Activates the magnetos by opening the grounding or 'p-lead' circuit; with the p-lead ungrounded the magneto is free to send its high-voltage output to the spark plugs. In most aircraft the ignition switch also applies power to the starter motor during engine start. In piston aircraft engines, the battery does not generate the spark for combustion. This is accomplished using devices called magnetos. Magnetos are connected to the engine by gearing. When the crankshaft turns, it turns the magnetos which mechanically generate voltage for spark. In the event of an electrical failure, the engine will continue to run. The Ignition Switch has the following positions:
    - Off - Both magneto p-leads are connected to electrical ground. This disables both magnetos, no spark is produced.
    - Right - The left magneto p-lead is grounded, and the right is open. This disables the left magneto and enables the right magneto only.

  - Left - The right magneto p-lead is grounded, and the left is open. This disables the right magneto and enables the left magneto only.
  - Both - This is the normal operating configuration, both p-leads are open, enabling both magnetos.
  - Start - The pinion gear on the starter motor is engaged with the flywheel and the starter motor runs to turn the engine over. In most cases, only the left magneto is active (the right p-lead is grounded) due to timing differences between the magnetos at low RPMs.
- Tachometer - A gauge to indicate engine speed in revolutions per minute (RPM) or percentage of maximum.
- Manifold Pressure (MP) Gauge - Indicates the absolute pressure in the intake manifold.
- Oil Temperature Gauge - Indicates the engine oil temperature.
- Oil Pressure Gauge - Indicates the supply pressure of the engine lubricant.
- Exhaust Gas Temperature (EGT) Gauge - Indicates the temperature of the exhaust gas just after combustion. Used to set the fuel/air mixture (leaning) correctly.
- Cylinder Head Temperature (CHT) Gauge - Indicates the temperature of at least one of the cylinder heads. Used to set the fuel/air mixture.
- Carburetor Heat Control - Controls the application of heat to the carburetor venturi area to remove or prevent the formation of ice in the throat of the carburetor as well as bypassing the air filter in case of impact icing.
- Alternate Air - Bypasses the air filter on a fuel-injected engine.

## Fuel

- Fuel Primer Pump - A manual pump to add a small amount of fuel at the cylinder intakes to assist in starting a cold engine. Fuel injected engines do not have this control. For fuel injected engines, a fuel boost pump is used to prime the engine prior to start.
- Fuel Quantity Gauge - Indicates the amount of fuel remaining in the identified tank. One per fuel tank. Some aircraft use a single gauge for all tanks, with a selector switch that can be turned to select the tank one wishes to have displayed on the shared gauge, including a setting to show the total fuel in all tanks. An example of switch settings could be “Left, Right, Fuselage, Total”. This saves room on the instrument panel by negating the need for four different dedicated fuel gauges.
- Fuel Select Valve - Connects the fuel flow from the selected tank to the engine.

If the aircraft is equipped with a fuel pump:

- Fuel Pressure Gauge - Indicates the supply pressure of fuel to the carburetor (or in the case of a fuel injected engine, to the fuel controller.)
- Fuel Boost Pump Switch - Controls the operation of the auxiliary electric fuel pump to provide fuel to the engine before it starts or in case of failure of the engine powered fuel pump.

Some large airplanes have a fuel system that allows the flight crew to jettison or dump the fuel. When operated, the boost pumps in the fuel tanks pump the fuel to the dump chutes or jettison nozzles and overboard to atmosphere.

## Propeller

If the aircraft is equipped with adjustable-pitch or constant-speed propeller(s):

- Propeller Control - Used to set the desired propeller speed. Once the pilot has set the desired propeller speed, the propeller governor maintains that propeller speed by adjusting the pitch of the propeller blades, using the engine's oil pressure to move a hydraulic piston in the propeller hub.
- Manifold Pressure Gauge - Indicates the (absolute) pressure in the engine's intake manifold. When the engine is running normally, there is a good correlation between the intake manifold pressure and the torque the engine is developing.

## Cowl

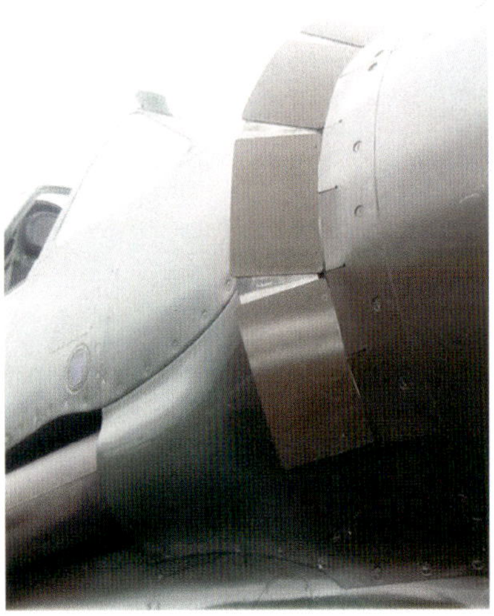

Front view of open cowl flaps

Rear view of open cowl flaps

If the aircraft is equipped with adjustable Cowl Flaps:

- Cowl Flap Position Control - Cowl Flaps are opened during high power/low airspeed operations like takeoff to maximize the volume of cooling airflow over the engine's cooling fins.
- Cylinder Head Temperature Gauge - Indicates the temperature of all cylinder heads or on a single CHT system, the hottest head. A Cylinder Head Temperature Gauge has a much shorter response time than the oil temperature gauge, so it can alert the pilot to a developing cooling issue more quickly. Engine overheating may be caused by:

1. Running too long at a high power setting.
2. Poor leaning technique.

3. Restricting the volume of cooling airflow too much.
4. Insufficient delivery of lubricating oil to the engine's moving parts.

In this conclusion, there are nuances to engine operations, but in general, if the pilot wants the aircraft to go faster the throttle is advanced; to slow down the throttle is retarded.

But the RPMs stay the same because the propeller control hasn't been moved. How does *that* work?

Simple. The propeller blades automatically change pitch, taking a bigger bite out of the air to go faster or less bite to go slower.

You'll notice that in none of this was there anything remotely resembling anything like a car's transmission.

## Full Authority Digital Engine Control (FADEC)

A sophisticated ignition: full authority digital engine control (FADEC) is a memory mapped microprocessor controlled ignition with sensors measuring MAP, CHT, EGT, RPM, atmospheric and oil pressure. It controls fuel injection and ignition timing to fully optimize the power produced by the engine.

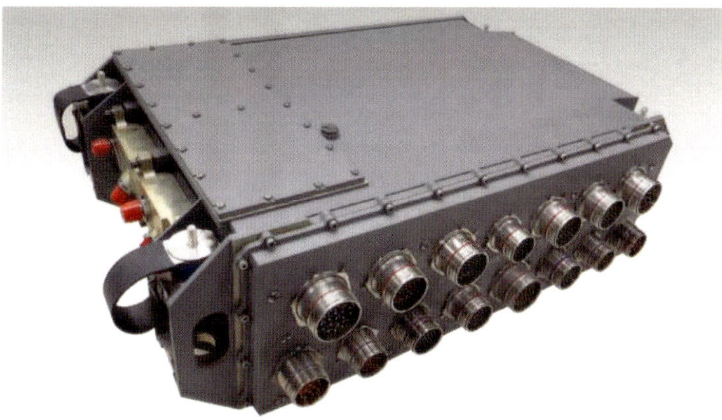

These systems add considerable complexity and additional wiring to the original electrical system as they need (among other things) a separate backup power supply from second battery. This secondary battery must be kept fully charged during flight through a special schottky diode (low voltage drop type) and must be monitored for charging. Due to the added weight of the parts of the FADEC the weight and balance of the aircraft must be recalculated after installation.

Cockpit controls are added and you will have extra switches for a fuel pump, primary and secondary FADEC power supply and fuses to protect these systems. There is no mixture control as the FADEC controls the amount of mixture injection and timing.

## Computer Controlled

The engine control unit (ECU) uses a 3D memory map to control the injector for the right amount of fuel and takes into account different ambient circumstances as outside air temperature (OAT) and air pressure (density) with regard to RPM and throttle setting and regulates this several times per second. Mixture control is automatic as the ECU senses barometric pressure and compensates the amount of fuel injected while the aircraft climbs.

## Ignition

The ignition timing of the spark plugs (spark advance) is also regulated depending on RPM, throttle setting for every load on the engine. This variable igniting timing results in quicker engine starts and smoother operation under variable loads.

Starting a FADEC engine is no more exciting than pressing the start button and it runs, no more choking, priming and endless battery draining starts. The ECU takes all variables into account and retards ignition and regulates fuel per cylinder for easy starts, just like any modern car these days.

## Fuel Injection

With this system the carburetor is finally eliminated. This is a big advantage as the change for carburetor icing is then nil. Carb ice develops when the temperature drops (sometimes in excess of 25°C) due to evaporation of the fuel and decrease in air pressure in the venturi throat. Should water vapor be present in the air, it condenses and freezes on the throttle valve and in the venturi, restricting airflow.

This can be resolve by applying carburetor heat. Intake air is rerouted over the hot exhaust and fed to the carburetor. The result is that the ice melts and the engine loses a bit of power, as hot air is less dens and the mixture will be richer.

The inlet air to the exhaust muff is not filtered so the pilot must switch carb heat off on short final, just in case he needs to make a go-around with full engine power and secondly the air close to the ground contains fine dust and sand which increases engine wear in the pistons and cylinders.

## Vapour Lock

Due to the high pressure fuel system, things as vapour lock are a non event. Even hot starting a fuel injected engine is no problem at all, the higher fuel pressure will remove any air pockets in the fuel lines.

## Fuel Savings

With a FADEC the engine receives the right amount of fuel per cylinder and the spark plugs are ignited at the right time regarding RPM, throttle setting, ambient temperature and pressure. This results in fuel savings up to 15%, an easier starting and smoother running engine.

## References


- "Demon UAV jets into history by flying without flaps". Metro.co.uk. London: Associated Newspapers Limited. 28 September 2010. Retrieved 29 September 2010
- Crouch, Tom (1982). Blériot XI, The Story of a Classic Aircraft. Smithsonian Institution Press. pp. 21 & 22. ISBN 0-87474-345-1
- Langewiesche, Wolfgang. Stick and Rudder: An Explanation of the Art of Flying, McGraw-Hill Professional, 1990, ISBN 0-07-036240-8, ISBN 978-0-07-036240-6
- Aerospace Engineering. London: Mechanical Engineering Publications. 224 (4): 355–363
- "Air Brakes for Planes Greatly Reduce the Landing Speed". 122(1). Popular Science. January 1933: 18. Retrieved 25 October 2012
- Smith, Apollo M. O. (1975). "High-Lift Aerodynamics" (PDF). Journal of Aircraft. 12 (6): 518–523. doi:10.2514/3.59830. ISSN 0021-8669. Archived from the original (PDF) on 7 July 2011. Retrieved 12 July 2011
- Crane, Dale: Dictionary of Aeronautical Terms, third edition, page 563. Aviation Supplies & Academics, 1997. ISBN 1-56027-287-2
- "Detailed Comparison of the Cessna TTx and Cirrus SR22T G5 | High Performance Aviation, LLC". High Performance Aviation. Retrieved 2018-03-13
- Crane, Dale: Dictionary of Aeronautical Terms, third edition, page 132. Aviation Supplies & Academics, 1997. ISBN 1-56027-287-2
- Crouch, Tom (1982). Blériot XI, The Story of a Classic Aircraft. Smithsonian Institution Press. pp. 21–22. ISBN 0-87474-345-1
- Gibbs-Smith, C.H.. Aviation: An Historical Survey From Its Origins To The End Of The Second World War, Science Museum, 1960 [2000], p.54, ISBN 1-900747-52-9, ISBN 978-1-900747-52-3. Retrieved 4 March 2013
- P John (2010). "The flapless air vehicle integrated industrial research (FLAVIIR) programme in aeronautical engineering". Proceedings of the Institution of Mechanical Engineers, Part G: Journal of
- Reitsch, Hanna (April 1997) [1955]. The Sky My Kingdom: Memoirs of the Famous German WWII Test-Pilot (Greenhill Military Paperback). Stackpole Books. p. 108. ISBN 9781853672620

5

# Aircraft Parts, Design and Structures

Science and technology have undergone rapid developments in the past decade, which has resulted in innovation in aircraft parts, design and structures. Many significant topics in the field of aircraft designing such as airframe, fuselage, wing, propeller, landing gear, etc. have been extensively detailed in this chapter.

## Airframe

Airframe, basic structure of an airplane or spacecraft excluding its power plant and instrumentation; its principal components thus include the wings, fuselage, tail assembly, and landing gear. The airframe is designed to withstand all aerodynamic forces as well as the stresses imposed by the weight of the fuel, crew, and payload.

Most airframes of early airplanes consisted of a fuselage of truss design constructed of narrow hardwood boards or steel tubing and braced with wires. This basic framework supported the wing structure, which was composed of spars with ribbing. Both the fuselage and wings were covered by a skin of cotton fabric. Airframe construction was radically improved during the 1930s. The aerodynamically contoured fuselage shell characteristic of all modern aircraft was introduced at this time, and high-strength, lightweight metals (chiefly aluminum alloys, magnesium, and some stainless steel and titanium) replaced wood and fabric throughout the airframe.

### Modern Era

Rough interior of a Boeing 747 airframe

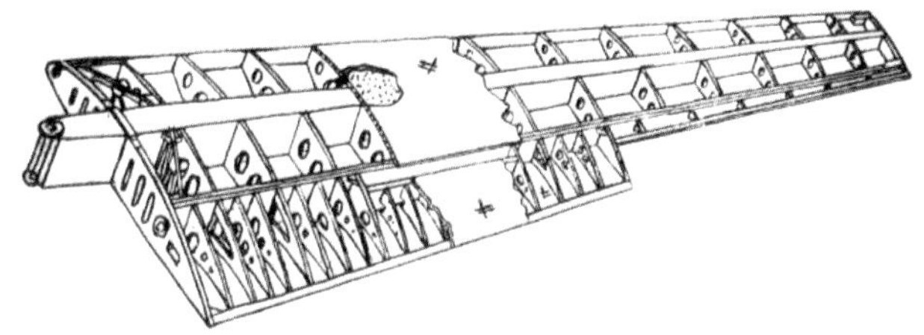

Wing structure with ribs and one spar

Airbus and Boeing are the dominant assemblers of large jet airliners while ATR, Bombardier and Embraer lead the regional airliner market; many manufacturers produce airframe components.

The vertical stabilizer of the Airbus A310-300, first flown in 1985, was the first carbon-fiber primary structure used in a commercial aircraft; composites are increasingly used since in Airbus airliners: the horizontal stabilizer of the A320 in 1987 and A330/A340 in 1994, and the center wing-box and aft fuselage of the A380 in 2005.

The Cirrus SR20, type certificated in 1998, was the first widely produced general aviation aircraft manufactured with all-composite construction, followed by several other light aircraft in the 2000s.

The Boeing 787, first flown in 2009, was the first commercial aircraft with 50% of its structure weight made of carbon-fiber composites, along 20% Aluminum and 15% titanium: the material allows for a lower-drag, higher wing aspect ratio and higher cabin pressurization; the competing Airbus A350, flown in 2013, is 53% carbon-fiber by structure weight. It has a one-piece carbon fiber fuselage, said to replace "1,200 sheets of aluminum and 40,000 rivets."

The 2013 Bombardier CSeries have a dry-fiber resin transfer infusion wing with a lightweight aluminium-lithium alloy fuselage for damage resistance and repairability, a combination which could be used for future narrow-body aircraft. In February 2017, Airbus installed a 3D printing machine for titanium aircraft structural parts using electron beam additive manufacturing from Sciaky, Inc.

## Safety

DH106 Comet 3 G-ANLO demonstrating at the 1954 Farnborough Airshow

Airframe production has become an exacting process. Manufacturers operate under strict quality control and government regulations. Departures from established standards become objects of major concern.

A landmark in aeronautical design, the world's first jet airliner, the de Havilland Comet, first flew in 1949. Early models suffered from catastrophic airframe metal fatigue, causing a series of widely publicised accidents. The Royal Aircraft Establishment investigation at Farnborough Airport founded the science of aircraft crash reconstruction. After 3000 pressurisation cycles in a specially constructed pressure chamber, airframe failure was found to be due to stress concentration, a consequence of the square shaped windows. The windows had been engineered to be glued and riveted, but had been punch riveted only. Unlike drill riveting, the imperfect nature of the hole created by punch riveting may cause the start of fatigue cracks around the rivet.

The Lockheed L-188 Electra turboprop, first flown in 1957 became a costly lesson in controlling oscillation and planning around metal fatigue. Its 1959 crash of Braniff Flight 542 showed the difficulties that the airframe industry and its airline customers can experience when adopting new technology.

The incident bears comparison with the Airbus A300 crash on takeoff of the American Airlines Flight 587 in 2001, after its vertical stabilizer broke away from the fuselage, called attention to operation, maintenance and design issues involving composite materials that are used in many recent airframes. The A300 had experienced other structural problems but none of this magnitude.

## Wing

The aircraft wing has transformed from the wooden and fabric twin-wing set up of the Wright brothers' Flyer, to the composite materials used in the latest models coming off the production line today from the likes of Boeing and Airbus. In fact, Orville Wright's first flight, lasting 12 seconds, travelled 120ft – less than the wing span of a Boeing 747. The basic forces acting on an aircraft are thrust, drag, lift and weight. OEMs are today placing a major focus on trying to reduce drag and weight across all components, in an effort to improve performance and efficiency.

(Credit: John T Daniels)

While the basic aerofoil shapes laid out in the 1930s by NASA (and its predecessor NACA) have remained, the fundamental platform for generating the lift component, there have been huge advances in both the materials and manufacturing processes used.

Efficiency is at the front of airline executives' minds. As an increasingly demanding customer base calls for cheaper flights, cost effectiveness in both the operation and manufacturing of the aircraft is crucial for airlines to maintain profit margins. As airlines constantly push their manufacturing suppliers to deliver more efficient aircraft, OEMs are turning to technology to drive improvements in wing design and manufacture, focusing on improvements in aerodynamics and reductions in component weight.

## Airplane Wings

The sole purpose of an aircraft wing is to generate the force of lift required to counter the weight of the airplane. Photo Credit: Casey Hussein Bisson

Of the four primary forces that enable an airplane to fly, liftis fundamental to flight. All vertical movement of the aircraft is a consequence of a relative imbalance between the weight of the aircraft and the force of lift produced by the airplane wings.

The wings of an aircraft are airfoils designed to create this force of vertical motion. Any surface that alters the airflow to the advantage of a produced force in a particular direction is termed as an airfoil. Airplane wings, designed as airfoils, achieve this by interacting with the remote airflow to produce the desired lift.

## Aerodynamics

Condensation in the low pressure region over the wing of an Airbus A340, passing through humid air

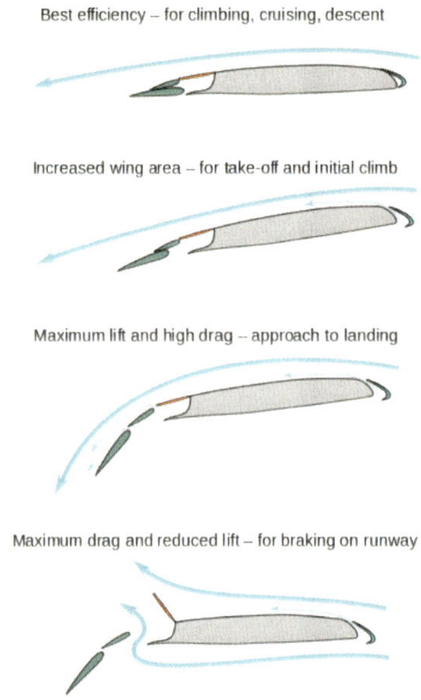


Flaps (green) are used in various configurations to increase the wing area and to increase the lift. In conjunction with spoilers (red), flaps maximize drag and minimize lift during the landing roll.

The design and analysis of the wings of aircraft is one of the principal applications of the science of aerodynamics, which is a branch of fluid mechanics. The properties of the airflow around any moving object can – in principle – be found by solving the Navier-Stokes equations of fluid dynamics. However, except for simple geometries these equations are notoriously difficult to solve. Fortunately, simpler explanations can be described.

For a wing to produce "lift", it must be oriented at a suitable angle of attack relative to the flow of air past the wing. When this occurs the wing deflects the airflow downwards, "turning" the air as it passes the wing. Since the wing exerts a force on the air to change its direction, the air must exert a force on the wing, equal in size but opposite in direction. This force manifests itself as differing air pressures at different points on the surface of the wing.

A region of lower-than-normal air pressure is generated over the top surface of the wing, with a higher pressure on the bottom of the wing. These air pressure differences can be either measured directly using instrumentation, or can be calculated from the airspeed distribution using basic physical principles—including Bernoulli's principle, which relates changes in air speed to changes in air pressure.

The lower air pressure on the top of the wing generates a smaller downward force on the top of the wing than the upward force generated by the higher air pressure on the bottom of the wing. Hence, a net upward force acts on the wing. This force is called the "lift" generated by the wing.

The different velocities of the air passing by the wing, the air pressure differences, the change in direction of the airflow, and the lift on the wing are intrinsically one phenomenon. It is, therefore, possible to calculate lift from any of the other three. For example, the lift can be calculated from the pressure differences, or from different velocities of the air above and below the wing, or from the total momentum change of the deflected air. Fluid dynamics offers other approaches to solving these problems—and all produce the same answers if done correctly. Given a particular wing and its velocity through the air, debates over which mathematical approach is the *most convenient* to use can be mistaken by novices as differences of opinion about the basic principles of flight.

## Cross-sectional Shape

Wings with an asymmetrical cross section are the norm in subsonic flight. Wings with a symmetrical cross section can also generate lift by using a positive angle of attack to deflect air downward. Symmetrical airfoils have higher stalling speeds than cambered airfoils of the same wing area but are used in aerobatic aircraft as they provide practical performance whether the aircraft is upright or inverted. Another example comes from sailboats, where the sail is a thin membrane with no path-length difference between one side and the other.

For flight speeds near the speed of sound (transonic flight), airfoils with complex asymmetrical shapes are used to minimize the drastic increase in drag associated with airflow near the speed of sound. Such airfoils, called supercritical airfoils, are flat on top and curved on the bottom.

## Design Features

The wing of a landing BMI Airbus A319-100. The slats at its leading edge and the flaps at its trailing edge are extended.

Aircraft wings may feature some of the following:

- A rounded leading edge cross-section,
- A sharp trailing edge cross-section,
- Leading-edge devices such as slats, slots, or extensions,
- Trailing-edge devices such as flaps or flaperons (combination of flaps and ailerons),
- Winglets to keep wingtip vortices from increasing drag and decreasing lift,
- Dihedral, or a positive wing angle to the horizontal, increases *spiral stability* around the roll axis, whereas *anhedral*, or a negative wing angle to the horizontal, decreases spiral stability.

Aircraft wings may have various devices, such as flaps or slats that the pilot uses to modify the shape and surface area of the wing to change its operating characteristics in flight.

- Ailerons (usually near the wingtips) to roll the aircraft clockwise or counterclockwise about its long axis.
- Spoilers on the upper surface to disrupt the lift and to provide additional traction to an aircraft that has just landed but is still moving.
- Vortex generators to help prevent flow separation in transonic flow.
- Wing fences to keep flow attached to the wing by stopping boundary layer separation from spreading roll direction.
- Folding wings allow more aircraft storage in the confined space of the hangar deck of an aircraft carrier.
- Variable-sweep wing or "swing wings" that allow outstretched wings during low-speed flight (i.e., take-off and landing) and swept back wings for high-speed flight (including supersonic flight), such as in the F-111 Aardvark, the F-14 Tomcat, the Panavia Tornado, the MiG-23, the MiG-27, the Tu-160 and the B-1B Lancer warplanes.

### Applications

Besides fixed-wing aircraft, applications for wing shapes include:

- Hang gliders, which use wings ranging from fully flexible (paragliders, gliding parachutes), flexible (framed sail wings), to rigid.
- Kites, which use a variety of lifting surfaces.
- Flying model airplanes.
- Helicopters, which use a rotating wing with a variable pitch angle to provide directional forces.
- Propellers, whose blades generate lift for propulsion.
- The NASA Space Shuttle, which uses its wings only to glide during its descent to a runway. These types of aircraft are called spaceplanes.
- Some racing cars, especially Formula One cars, which use upside-down wings (or *airfoils*) to provide greater traction at high speeds.
- Sailboats, which use sails as vertical wings with variable fullness and direction to move across water.

## Fuselage

The fuselage, or body of the airplane, is a long hollow tube which holds all the pieces of an airplane together. The fuselage is hollow to reduce weight. As with most other parts of the airplane, the shape of the fuselage is normally determined by the mission of the aircraft. A supersonic fighter plane has a very slender, streamlined fuselage to reduce the drag associated with high speed flight. An airliner has a wider fuselage to carry the maximum number of passengers. On an airliner, the pilots sit in a cockpit at the front of the fuselage. Passengers and cargo are carried in the rear of the fuselage and the fuel is usually stored in the wings. For a fighter plane, the cockpit is normally on top of the fuselage, weapons are carried on the wings, and the engines and fuel are placed at the rear of the fuselage.

The weight of an aircraft is distributed all along the aircraft. The fuselage, along with the passengers and cargo, contribute a significant portion of the weight of an aircraft. The center of gravity of the aircraft is the average location of the weight and it is usually located inside the fuselage. In flight, the aircraft rotates around the center of gravity because of torques generated by the elevator, rudder, and ailerons. The fuselage must be designed with enough strength to withstand these torques.

Most people who have ever travelled in a commercial jet will realise that it is the fuselage of the aircraft that protects them from the world outside. Essentially, an aeroplane's fuselage is the sheet metal covering on the outside of the vehicle that allows the internal cabin to be pressurised such that its passengers and crew are not exposed to the low pressure outside which would be fatal given the high altitudes modern jets fly at. However, the highly engineered metalwork that makes up a plane's fuselage provides much more than a mere outer layer. Other than being the 'skin' of an aircraft, what it is that makes it so important to the safe running of a modern jet?

## Structural Integrity

It is a mistake to think of the metal fuselage of a passenger or freight jet as only providing a bubble of safety inside for a plane's occupants. Yes, it performs this function, but modern aircraft use their fuselage to provide a great deal of structural strength as well as an external covering. All of the principle elements of the internal section of a jet plane are attached to it. Therefore, the rolled metalwork that makes the fuselage must be processed with a great deal of attention to detail if it is to be able to connect properly with things like the plane's floor and luggage compartment. So-called monocoque design planes rely on the shell of the aircraft to provide structural integrity. A semi-monocoque fuselage has a reinforcing framework behind it, but the principle remains the same – the fuselage's structure is integral to the design of the plane.

## Metal Fuselage

As mentioned, sheet metalwork is commonly used these days to form the basis of a plane's fuselage. Rolled into circular sections, lightweight but extremely durable metals are chosen for the job. A fine example of this would be aluminium which many modern planes are made from. In the early days of aviation, wood or even canvas might have been used but neither provided the required durability. Remember that fuselage must withstand stresses cause from a plane's own jet engines forcing it through the air, but also from physical objects it might come into contact with, such as hail or bird strikes.

## Recyclable Fuselage

Contemporary aircraft designers must also consider the materials they use for fuselage these days. Some aircraft designs rely on new materials, like graphite epoxy polymers. However, these cannot always be satisfactorily recycled into new products once the fuselage has come to the end of its working life. This is in contrast to sheet metal materials, like titanium and aluminium, which are often reworked into new components, including new sections of fuselage from time-to-time.

## Types of Structures

### Truss Structure

Piper PA-18 welded tube truss fuselage structure

This type of structure is still in use in many lightweight aircraft using welded steel tube trusses. A box truss fuselage structure can also be built out of wood—often covered with plywood. Simple box structures may be rounded by the addition of supported lightweight stringers, allowing the fabric covering to form a more aerodynamic shape, or one more pleasing to the eye.

### Geodesic Construction

Geodesic airframe fuselage structure is exposed by battle damage

Geodesic structural elements were used by Barnes Wallis for British Vickers between the wars and into World War II to form the whole of the fuselage, including its aerodynamic shape. In this type of construction multiple flat strip stringers are wound about the formers in opposite spiral directions, forming a basket-like appearance. This proved to be light, strong, and rigid and had the advantage of being made almost entirely of wood. A similar construction using aluminum alloy was used in the Vickers Warwick with less materials than would be required for other structural types.

The geodesic structure is also redundant and so can survive localized damage without catastrophic failure. A fabric covering over the structure completed the aerodynamic shell. The logical evolution of this is the creation of fuselages using molded plywood, in which multiple sheets are laid with the grain in differing directions to give the monocoque type below.

## Monocoque Shell

The Van's Aircraft RV-7 of semi-monocoque construction.

In this method, the exterior surface of the fuselage is also the primary structure. A typical early form of this was built using molded plywood, where the layers of plywood are formed over a "plug" or within a mold. A later form of this structure uses fiberglass cloth impregnated with polyester or epoxy resin, instead of plywood, as the skin. A simple form of this used in some amateur-built aircraft uses rigid expanded foam plastic as the core, with a fiberglass covering, eliminating the necessity of fabricating molds, but requiring more effort in finishing. An example of a larger molded plywood aircraft is the de Havilland Mosquito fighter/light bomber of World War II. No plywood-skin fuselage is truly monocoque, since stiffening elements are incorporated into the structure to carry concentrated loads that would otherwise buckle the thin skin. The use of molded fiberglass using negative ("female") molds (which give a nearly finished product) is prevalent in the series production of many modern sailplanes. The use of molded composites for fuselage structures is being extended to large passenger aircraft such as the Boeing 787 *Dreamliner* (using pressure-molding on female molds).

## Semi-monocoque

This is the preferred method of constructing an all-aluminum fuselage. First, a series of frames in the shape of the fuselage cross sections are held in position on a rigid fixture. These frames are then joined with lightweight longitudinal elements called stringers. These are in turn covered with a skin of sheet aluminum, attached by riveting or by bonding with special adhesives. The fixture is then disassembled and removed from the completed fuselage shell, which is then fitted out with wiring, controls, and interior equipment such as seats and luggage bins. Most modern large aircraft are built using this technique, but use several large sections constructed in this fashion which are then joined with fasteners to form the complete fuselage. As the accuracy of the final product is determined largely by the costly fixture, this form is suitable for series production, where a large number of identical aircraft are to be produced. Early examples of this type include the Douglas

Aircraft DC-2 and DC-3 civil aircraft and the Boeing B-17 Flying Fortress. Most metal light aircraft are constructed using this process.

Sectioned fuselage showing frames, stringers and skin all made of aluminium

Both monocoque and semi-monocoque are referred to as "stressed skin" structures as all or a portion of the external load (i.e. from wings and empennage, and from discrete masses such as the engine) is taken by the surface covering. In addition, all the load from internal pressurization is carried (as *skin tension*) by the external skin.

The proportioning of loads between the components is a design choice dictated largely by the dimensions, strength, and elasticity of the components available for construction and whether or not a design is intended to be "self jigging", not requiring a complete fixture for alignment.

## Materials

Inside view of the wooden, fabric-covered fuselage of a Fisher FP-202.

Early aircraft were constructed of wood frames covered in fabric. As monoplanes became popular, metal frames improved the strength, which eventually led to all-metal aircraft with metal covering all surfaces. Some modern aircraft are constructed with composite materials for major control surfaces, wings, or the entire fuselage such as the Boeing 787. On the 787, it makes possible higher pressurization levels and larger windows for passenger comfort as well as lower weight to reduce operating costs. The Boeing 787 weighs 1500 lb less than if it were an all-aluminum assembly.

### Windows

Cockpit windshields on the Airbus A320 must withstand bird strikes up to 350 kt and are made of chemically strengthened glass. They are usually composed of three layers or plies, of glass or plastic : the inner two are 8 mm (0.3 in.) thick each and are structural, while the outer ply, about 3 mm thick, is a barrier against foreign object damage and abrasion, with often a hydrophobic coating. It must prevent fogging inside the cabin and de-ice from −50 °C (−58 °F), this was previously done with thin wires similar to a rear car window, but now is a transparent, nanometers thick coating of indium tin oxide siting between plies, electrically conductive and thus transmitting heat. Curved glass improves aerodynamics but sight criteria also needs larger panes. A cockpit windshield is composed of 4–6 panels, 35 kg (77 lb.) each on an Airbus A320. In its lifetime, an average aircraft goes through three or four windshields, and the market is shared evenly between OEM and higher margins aftermarket.

Cabin windows, made from much lighter than glass stretched acrylic glass, consists of multiple panes: an outer one built to support four times the maximum cabin pressure, an inner one for redundancy and a scratch pane near the passenger. Acrylic is susceptible to crazing : a network of fine cracks appears but can be polished to restore optical transparency, removal and polishing typically undergo every 2–3 years for uncoated windows.

### Wing Integration

"Flying wing" aircraft, such as the Northrop YB-49 Flying Wing and the Northrop B-2 Spirit bomber have no separate fuselage; instead what would be the fuselage is a thickened portion of the wing structure.

Conversely there have been a small number of aircraft designs which have no separate wing, but use the fuselage to generate lift. Examples include National Aeronautics and Space Administration's experimental lifting body designs and the Vought XF5U-1 Flying Flapjack.

A blended wing body can be considered a mixture of the above. It carries the useful load in a fuselage producing lift. A modern example is Boeing X-48. One of the earliest aircraft using this design approach is Burnelli CBY-3, which fuselage was airfoil shaped to produce lift.

## Landing Gear

Just as a bird eventually needs to stretch its legs, so too does an airplane require some form of landing gear. The gear in turn requires an undercarriage, or a structure that supports the plane's weight on the ground.

The Wright brothers' 1903 flyer depended on simple wooden skids for landing in the sand. Other more modern craft to possess landing skids include the German Messerschmitt ME 163 Komet, a World

War II rocket-propelled interceptor, and the U.S. Air Force's X-15, an experimental, high-speed 1960s jet. Along similar lines, some aircraft boast floats or skis for landing on water, snow or ice.

Landing Gear system is one of the critical subsystems of an aircraft and is often configured along with the aircraft structure because of its substantial influence on the aircraft structural configuration itself. Landing gear detail design is taken up early in the aircraft design cycle due to its long product development cycle time. The need to design landing gear with minimum weight, minimum volume, reduced life cycle cost, and short development cycle time, poses many challenges to landing gear designers and practitioners. These challenges have to be met by employing advanced technologies, materials, analysis methods, processes and production methods. Various design and analysis tools have been developed over the years and new ones are still being developed.

The purpose of the landing gear in an aircraft is to provide a suspension system during taxi, take-off and landing. It is designed to absorb and dissipate the kinetic energy of landing impact, thereby reducing the impact loads transmitted to the airframe. The landing gear also facilitates braking of the aircraft using a wheel braking system and provides directional control of the aircraft on ground using a wheel steering system. It is often made retractable to minimize the aerodynamic drag on the aircraft while flying.

The landing gear design takes into account various requirements of strength, stability, stiffness, ground clearance, control and damping under all possible ground attitudes of the aircraft. These requirements are stipulated by the Airworthiness Regulations to meet operational requirements and safety. The landing gear should occupy minimum volume in order to reduce the stowage space requirement in the aircraft. Further, weight should be at minimum to increase the performance of the aircraft. The service life of the landing gears should be same as that of the aircraft.

When you think of landing gear, however, you probably think of the wheeled variety. The actual wheels involved have ranged over the vast spectrum of aviation designs. Some early landing gear resembled bicycle wheels while larger aircraft often feature bogie landing gear that employ sets of four or more wheels on each brace. During the 1950s, the U.S. Air Force even experimented with tank-style tracked landing gear for the enormous six-engine Convair B-36 Peacemaker.

Regardless of the type of wheel employed, such landing gear are typically arranged in one of two arrangements. First there's the conventional undercarriage with two front wheels and one smaller tail wheel or skid. You can spot this arrangement, also known as a taildragger undercarriage, on older prop-driven aircraft. Most modern planes use a tricycle undercarriage, in which the smaller wheel is positioned at the front of an aircraft.

Variations on these two basic themes are numerous, with additional wheels added depending on the particular demands of a given aircraft. The Lockheed U-2, for instance, features a tandem design with two fuselage wheels running down the middle and supporting wheels on each wing for balance. Many modern aircraft feature retractable landing gear, which pull up into fuselage during flight, but others still feature fixed landing gear that remain extended all the time.

## Aircraft Landing Gear

Aircraft landing gear usually includes wheels equipped with simple shock absorbers, or more advanced air/oil oleo struts, for runway and rough terrain landing. Some aircraft are equipped with skis for snow or floats for water, and/or skids or pontoons (helicopters).

It represents 2.5 to 5% of the MTOW and 1.5 to 1.75% of the aircraft cost but 20% of the airframe direct maintenance cost; each wheel can support up to 30 t (66,000 lb), reach over 300 km/h, roll up to 500,000 km (310,000 mi) ; it has a 20,000 hours time between overhaul and a 60,000 hours or 20 years life time. The undercarriage is typically 4–5% of the takeoff mass and can even reach 7%.

## Gear Arrangements

Wheeled undercarriages normally come in two types:

- *Conventional or "taildragger" undercarriage*, where there are two main wheels towards the front of the aircraft and a single, much smaller, wheel or skid at the rear;
- *Tricycle undercarriage* where there are two main wheels (or wheel assemblies) under the wings and a third smaller wheel in the nose.

Conventional "taildragger" undercarriage (SAN Jodel D.140 Mousquetaire)

Tricycle undercarriage (Mooney M20J)

The taildragger arrangement was common during the early propeller era, as it allows more room for propeller clearance. Most modern aircraft have tricycle undercarriages. Taildraggers are considered harder to land and take off (because the arrangement is usually *unstable*, that is, a small deviation from straight-line travel will tend to increase rather than correct itself), and usually require special pilot training. Sometimes a small tail wheel or skid is added to aircraft with tricycle undercarriage, in case of tail strikes during take-off. Concorde, for instance, had a retractable tail "bumper" wheel, as delta winged aircraft need a high angle when taking off. Both Boeing's largest WW II bomber, the B-29 Superfortress and the 1960s-introduction Boeing 727 trijet airliner each have a retractable tail bumper. Some aircraft with retractable conventional landing gear have a fixed tailwheel, which generates minimal drag (since most of the airflow past the tailwheel has been blanketed by the fuselage) and even improves yaw stability in some cases. Another arrangement sometimes used is central main and nose gear with outriggers on the wings. This may be done where there is no convenient location on either side to attach the main undercarriage or to store it when retracted. Examples include the Lockheed U-2 spy plane and the Harrier Jump Jet.

Methods of Landing Gear Shock Absorption

Oleo (pressurized gas - oil) strut with torque links

Oleo strut with trailing link

Rubber discs in compression with trailing link, used on Mooney M20 series of aircraft

## Retractable Gear

To decrease drag in flight some undercarriages retract into the wings and/or fuselage with wheels flush against the surface or concealed behind doors; this is called *retractable gear*. If the wheels rest protruding and partially exposed to the airstream after being retracted, the system is called semi-retractable.

Most retraction systems are hydraulically operated, though some are electrically operated or even manually operated. This adds weight and complexity to the design. In retractable gear systems, the compartment where the wheels are stowed are called wheel wells, which may also diminish valuable cargo or fuel space.

Pilots confirming that their landing gear is down and locked refer to “three green” or “three in the green.”, a reference to the electrical indicator lights (or painted panels of mechanical indicator units) from the nosewheel/tailwheel and the two main gears. Red lights indicate the gear is in the up-locked position; amber lights indicate that the landing gear is in transit (neither down and locked nor fully retracted).

Multiple redundancies are usually provided to prevent a single failure from failing the entire landing gear extension process. Whether electrically or hydraulically operated, the landing gear can usually be powered from multiple sources. In case the power system fails, an emergency extension system is always available. This may take the form of a manually operated crank or pump, or a mechanical free-fall mechanism which disengages the uplocks and allows the landing gear to fall due to gravity. Some high-performance aircraft may even feature a pressurized-nitrogen back-up system.

The landing gear of a Boeing 767 retracting into the fuselage

A Boeing 737-700 with main undercarriage retracted in the wheel wells without landing gear doors

## Large Aircraft

As aircraft grow larger, they employ more wheels to cope with the increasing weights. The earliest "giant" aircraft ever placed in quantity production, the Zeppelin-Staaken R.VI German World War I long-range bomber of 1916, used a total of eighteen wheels for its undercarriage, split between two wheels on its nose gear struts, and a total of sixteen wheels on its main gear units — split into four side-by-side quartets each, two quartets of wheels per side — under each tandem engine nacelle, to support its loaded weight of almost 12 metric tons. Multiple "tandem wheels" on an aircraft — particularly for cargo aircraft, mounted to the fuselage lower sides as retractable main gear units on modern designs — were first seen during World War II, on the experimental German Arado Ar 232 cargo aircraft, which used a row of eleven "twinned" fixed wheel sets directly under the fuselage centerline to handle heavier loads while on the ground. Many of today's large cargo aircraft use this arrangement for their retractable main gear setups (usually mounted on the lower corners of the central fuselage structure). The Airbus A340-500/-600 has an additional four-wheel undercarriage bogie on the fuselage centerline, much like the twin-wheel unit in the same general location, used on later DC-10 and MD-11 airliners.

The Boeing 747 has five sets of wheels: a nosewheel assembly and four sets of four-wheel bogies. A set is located under each wing, and two inner sets are located in the fuselage, a little rearward of the outer bogies, adding up to a total of eighteen wheels and tires. The Airbus A380 also has a four-wheel bogie under each wing with two sets of six-wheel bogies under the fuselage.

The world's largest jet cargo aircraft, the Ukrainian Antonov An-225 has 4 wheels on the twin-strut nose gear units (as its smaller "stablemate", the Antonov An-124 also uses), and 28 main gear wheel/tire units, adding up to a total of 32 wheels and tires.

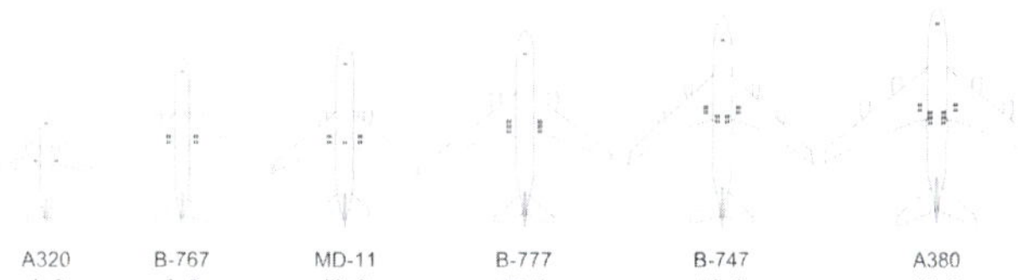


Wheel arrangements of large airliners

Oleo strut rear landing gear of an Antonov An-124 (24 wheel landing gear)

The A340-600 has an additional main undercarriage on the fuselage belly

## Nautical

Some aircraft have landing gear adapted to take off from and land on water.

A floatplane has landing gear comprising two or more streamlined floats.

A flying boat has a lower fuselage possessing the shape of a boat hull giving it buoyancy, usually with a "step" near the center of gravity to allow the aircraft to more easily break free of the water's surface for takeoff. Additional landing gear is often present, typically comprising wing-mounted floats, or more rarely, stub-wing like sponsons on the lower sides of the fuselage, with their lower surfaces even with the chines forming the longitudinal lower corners of a flying boat's lower hull contours.

Helicopters able to land on water may have floats or a hull.

An amphibious aircraft has landing gear for both land and water-based operation.

## Other types of Landing Gear

Experimental tracked gear on a B-36 Peacemaker

## Detachable Landing Gear

Some aircraft use wheels for takeoff and then jettison them soon afterwards for improved aerodynamic streamlining without the complexity, weight and space requirements of a retraction mechanism. In these cases, the wheels to be jettisoned are sometimes mounted onto axles that are part of a separate "dolly" (for main wheels only) or "trolley" (for a three-wheel set with a nosewheel) chassis. Landing is then accomplished on skids or similar other simple devices.

Historical examples include the "dolly"-using Messerschmitt Me 163 *Komet* rocket fighter, the Messerschmitt Me 321 *Gigant* troop glider, and the first eight "trolley"-using prototypes of the Arado Ar 234 jet reconnaissance bomber. The main disadvantage to using the takeoff dolly/trolley and landing skid(s) system on German World War II aircraft – intended for a sizable number of late-war German jet and rocket-powered military aircraft designs – was that aircraft would likely be scattered all over a military airfield after they had landed from a mission, and would be unable to taxi on their own to an appropriately hidden "dispersal" location, which could easily leave them vulnerable to being shot up by attacking Allied fighters. A related contemporary example are the wingtip support wheels ("pogos") on the Lockheed U-2 reconnaissance aircraft, which fall away after take-off and drop to earth; the aircraft then relies on titanium skids on the wingtips for landing.

## Helicopters

Light helicopters tend to use simple landing skids to save weight and cost. They include attachment points for wheels so that they can be moved for short distances on the ground. Skids are impractical for helicopters weighing more than four tons. Some high-speed machines have retractable wheels, but most use fixed wheels for their robustness, and to avoid the need for a retraction mechanism.

## Rearwards and Sideways Retraction

A Royal Air Force P-47 with its raked-forward main gear, and rearward-angled main wheel position (when retracted) indicated by the just-visible open wheel door.

Some main landing gear struts on World War II aircraft, in order to allow a single-leg main gear to more efficiently store the wheel within either the wing or an engine nacelle, rotated the single gear strut through a 90° angle during the rearwards-retraction sequence to allow the main wheel to rest "flat" above the lower end of the main gear strut, or flush within the wing or engine nacelles, when fully retracted. Examples are the Curtiss P-40, Vought F4U Corsair, Grumman F6F Hellcat, Messerschmitt Me 210 and Junkers Ju 88. The Aero Commander family of twin-engined business aircraft also shares this feature on the main gears, which retract aft into the ends of the engine nacelles. The rearward-retracting nosewheel strut on the Heinkel He 219 and the forward-retracting nose gear strut on the later Cessna Skymaster similarly rotated 90 degrees as they retracted.

On most World War II single-engined fighter aircraft (and even one German heavy bomber design) with sideways retracting main gear, the main gear that retracted into the wings was meant to be raked forward, towards the aircraft's nose in the "down" position for better ground handling, with a retracted position that placed the main wheels at some angle "behind" the main gear's attachment point to the airframe – this led to a complex angular geometry for setting up the "pintle" angles at the top ends of the struts for the retraction mechanism's axis of rotation, with some aircraft, like the P-47 Thunderbolt and Grumman Bearcat, even mandating that the main gear struts lengthen as they were extended down from the wings to assure proper ground clearance for their large four-bladed propellers. One exception to the need for this complexity in many WW II fighter aircraft was Japan's famous Zero fighter, whose main gear stayed at a perpendicular angle to the centerline of the aircraft when extended, as seen from the side.

## Tandem Layout

An unusual undercarriage configuration is found on the Hawker Siddeley Harrier, which has two mainwheels in line astern under the fuselage (called a bicycle or *tandem* layout) and a smaller wheel near the tip of each wing. On second generation Harriers, the wing is extended past the outrigger wheels to allow greater wing-mounted munition loads to be carried, or to permit wing-tip extensions to be bolted on for ferry flights.

Hawker Siddeley Harrier GR7 (ZG472). The two mainwheels are in line astern under the fuselage, with a smaller wheel on each wing

A multiple tandem layout was used on some military jet aircraft during the 1950s, pioneered by the Martin XB-51, and later used on such aircraft as the U-2, Myasishchev M-4, Yakovlev Yak-25, Yak-28, Sud Aviation Vautour, and the B-47 Stratojet because it allows room for a large internal bay between the main wheels. A variation of the multi tandem layout is also used on the B-52 Stratofortress which has four main wheel bogies (two forward and two aft) underneath the fuselage and a small outrigger wheel supporting each wing-tip. The B-52's landing gear is also unique in that all four pairs of main wheels can be steered. This allows the landing gear to line up with the runway and thus makes crosswind landings easier (using a technique called *crab landing*). Since tandem aircraft cannot rotate for takeoff, the forward gear must be long enough to give the wings the correct angle of attack during takeoff. During landing, the forward gear must not touch the runway first, otherwise the rear gear will slam down and cause the aircraft to bounce off the runway.

## Crosswind Landing Accommodation

The "castoring" main gear arrangement on a Blériot XI

One very early undercarriage arrangement that passively allowed for castoring during crosswind landings, unlike the "active" arrangement on the B-52, was pioneered on the Bleriot VIII design of 1908. It was later used in the much more famous Blériot XI Channel-crossing aircraft of 1909 and also copied in the earliest examples of the Etrich Taube. In this arrangement the main landing gear's shock absorption was taken up by a vertically sliding bungee cord-sprung upper member. The vertical post along which the upper member slid to take landing shocks also had its lower end as the rotation point for the forward end of the main wheel's suspension fork, allowing the main gear to pivot on moderate crosswind landings.

## "Kneeling" Gear

One of the very first aircraft to use a "kneeling" function in its undercarriage design was the World War II German Arado Ar 232 cargo/transport aircraft, produced in small numbers as both a twin-engined version, and one with four engines - both the nosegear, and the wing-mounted, inwards-retracting main landing gear were designed to have a "kneeling" function in their design to assist in loading/unloading cargo, and to also allow its unique, exposed fixed ventral fuselage-centreline set of eleven "twinned" auxiliary wheel sets to more firmly support the fuselage on soft ground, and to enable taxiing the aircraft over ditches and other ground obstacles.

Some early U.S. Navy jet fighters were equipped with "kneeling" nose gear consisting of small steerable auxiliary wheels on short struts located forward of the primary nose gear, allowing the aircraft to be taxied tail-high with the primary nose gear retracted. This feature was intended to enhance safety aboard aircraft carriers by redirecting the hot exhaust blast upwards, and to reduce hangar space requirements by enabling the aircraft to park with its nose underneath the tail of a similarly equipped jet. Kneeling gear was used on the North American FJ-1 Fury and on early versions of the McDonnell F2H Banshee, but was found to be of little use operationally, and was omitted from later Navy fighters.

The nosewheel gear systems of some large cargo jets, like the Antonov An-124 *Condor*, kneel to assist in loading and unloading of cargo using ramps through the forward, "tilt-up" hinged fuselage nose while stationary on the ground.

## Folding Gear

In order to save precious space, various folding and splayable landing gear designs have been created.

## Light Aircraft

For light aircraft a type of landing gear which is economical to produce is a simple wooden arch laminated from ash, as used on some homebuilt aircraft. A similar arched gear is often formed from spring steel. The Cessna Airmaster was among the first aircraft to use spring steel landing gear. The main advantage of such gear is that no other shock-absorbing device is needed; the deflecting leaf provides the shock absorption.

## Monowheel

To minimize drag, modern gliders usually have a single wheel, retractable or fixed, centered under the fuselage, which is referred to as *monowheel gear* or *monowheel landing gear*. Monowheel gear is also used on some powered aircraft, where drag reduction is a priority, such as the Europa XS. Much like the Me 163 rocket fighter, some gliders from prior to the Second World War used a take-off dolly that was jettisoned on take-off and then landed on a fixed skid. This configuration is necessarily accompanied with a taildragger.

A Schleicher ASG 29 glider shows its monowheel landing gear

## Tailsitter

Experimental tailsitter aircraft use landing gear located in their tails for VTOL operation.

A Convair XFY-1 Pogo showing its landing gear

## Ground Carriage

The idea behind a ground carriage is to leave the landing gear on the runway and not take it into the air, in order to reduce weight and drag. Examples include the "dolly" and "trolley" arrangements, respectively of the German Me 163B rocket fighter and Arado Ar 234A prototype jet recon-bomber designs of World War II, as their wheeled "ground carriages" were not usually allowed to either remain attached to the airframe, nor carried vary far away from the ground, during a normal takeoff procedure for either design.

Wing and fuselage undercarriages on a Boeing 747-400, shortly before landing

## Steering

There are several types of steering. Taildragger aircraft may be steered by rudder alone (depending upon the prop wash produced by the aircraft to turn it) with a freely pivoting tail wheel, or by a steering linkage with the tail wheel, or by *differential braking* (the use of independent brakes on opposite sides of the aircraft to turn the aircraft by slowing one side more sharply than the other). Aircraft with tricycle landing gear usually have a steering linkage with the nosewheel (especially in large aircraft), but some allow the nosewheel to pivot freely and use differential braking and/or the rudder to steer the aircraft, like the Cirrus SR22.

Some aircraft require that the pilot steer by using rudder pedals; others allow steering with the yoke or control stick. Some allow both. Still others have a separate control, called a *tiller*, used for steering on the ground exclusively.

Wheel-skis

## Rudder Steering

When an aircraft is steered on the ground exclusively using the rudder, turning the plane requires that a substantial airflow be moving past the rudder, which can be generated either by the forward motion of the aircraft or by thrust provided by the engines. Rudder steering requires considerable practice to use effectively. Although it requires air movement, it has the advantage of being independent of the landing gear, which makes it useful for aircraft equipped with fixed floats or skis.

## Direct Steering

The nose gear steering-wheel (tiller) is visible as a semi-circular wheel to the left of the yoke in this photo of a Boeing 727 cockpit

Some aircraft link the yoke, control stick, or rudder directly to the wheel used for steering. Manipulating these controls turns the steering wheel (the nose wheel for tricycle landing gear, and the tail wheel for taildraggers). The connection may be a firm one in which any movement of the controls turns the steering wheel (and vice versa), or it may be a soft one in which a spring-like mechanism twists the steering wheel but does not force it to turn. The former provides positive steering but makes it easier to skid the steering wheel; the latter provides softer steering (making it easy to overcontrol) but reduces the probability of skidding. Aircraft with retractable gear may disable the steering mechanism wholly or partially when the gear is retracted.

## Differential Braking

Differential braking depends on asymmetric application of the brakes on the main gear wheels to turn the aircraft. For this, the aircraft must be equipped with separate controls for the right and left brakes (usually on the rudder pedals). The nose or tail wheel usually is not equipped with brakes. Differential braking requires considerable skill. In aircraft with several methods of steering that include differential braking, differential braking may be avoided because of the wear it puts on the braking mechanisms. Differential braking has the advantage of being largely independent of any movement or skidding of the nose or tailwheel.

## Tiller Steering

A tiller in an aircraft is a small wheel or lever, sometimes accessible to one pilot and sometimes duplicated for both pilots, that controls the steering of the aircraft while it is on the ground. The tiller may be designed to work in combination with other controls such as the rudder or yoke. In large airliners, for example, the tiller is often used as the sole means of steering during taxi, and then the rudder is used to steer during takeoff and landing, so that both aerodynamic control surfaces and the landing gear can be controlled simultaneously when the aircraft is moving at aerodynamic speeds.

## Tires and Wheels

Two mechanics replacing a main landing gear wheel on a Lockheed P-3 Orion

Luftwaffe groundcrew servicing a Heinkel He 177A's main-gear wheels and tires

## Gear Loading

The choice of the main wheel tires is made on the basis of the static loading case. The total main gear load $F_m$ is calculated assuming that the aircraft is taxiing at low speed without braking:

$$F_m = \frac{l_n}{l_m + l_n} W.$$

where $W$ is the weight of the aircraft and $l_m$ and $l_n$ are the distance measured from the aircraft's center of gravity(cg) to the main and nose gear, respectively.

The choice of the nose wheel tires is based on the nose wheel load during braking at maximum effort:

$$F_n = \frac{l_m}{l_m + l_n}(W - L) + \frac{h_{cg}}{l_m + l_n}\left(\frac{a_x}{g} W - D + T\right).$$

where $L$ is the lift, $D$ is the drag, $T$ is the thrust, and $h_{cg}$ is the height of aircraft cg from the static groundline. Typical values for $\frac{a_x}{g}$ on dry concrete vary from 0.35 for a simple brake system to 0.45 for an automatic brake pressure control system. As both $L$ and $D$ are positive, the maximum nose gear load occurs at low speed. Reverse thrust decreases the nose gear load, and hence the condition $T = 0$ results in the maximum value:

$$F_n = \frac{l_m + h_{cg}(\frac{a_x}{g})}{l_m + l_n} W.$$

To ensure that the rated loads will not be exceeded in the static and braking conditions, a seven percent safety factor is used in the calculation of the applied loads.

## Inflation Pressure

Provided that the wheel load and configuration of the landing gear remain unchanged, the weight and volume of the tire will decrease with an increase in inflation pressure. From the flotation standpoint, a decrease in the tire contact area will induce a higher bearing stress on the pavement, thus eliminates certain airports from the aircraft's operational bases. Braking will also become less effective due to a reduction in the frictional force between the tires and the ground. In addition, the decrease in the size of the tire, and hence the size of the wheel, could pose a problem if internal brakes are to be fitted inside the wheel rims. The arguments against higher pressure are of such a nature that commercial operators generally prefer the lower pressures in order to maximize tire life and minimize runway stress. To prevent punctures from stones Philippine Airlines had to operate their Hawker Siddeley 748 aircraft with pressures as low as the tire manufacturer would permit. However, too low a pressure can lead to an accident as in the Nigeria Airways Flight 2120.

A rough general rule for required tire pressure is given by the manufacturer in their catalog.

Goodyear for example advises the pressure to be 4% higher than required for a given weight or as fraction of the rated static load and inflation.

Tires of many commercial aircraft are required to be filled with nitrogen or low-oxygen air to prevent the internal combustion of the tire which may result from overheating brakes producing volatile hydrocarbons from the tire lining.

### Stowaways in Aircraft Landing Gear

Unauthorized passengers have been known to stowaway on larger aircraft by climbing a landing gear strut and riding in the compartment meant for the wheels. There are extreme dangers to this practice, with numerous deaths reported. Dangers include a lack of oxygen at high altitude, temperatures well below freezing, crush injury or death from the gear retracting into its confined space, and falling out of the compartment during takeoff or landing.

## Propeller

A C-130 Super Hercules with scimitar propellers providing thrust for take-off

As opposed to jet airplanes, which are propelled via jet engines, many aircraft make use of propellers, which slice through the air to create thrust. Within an air mass, airplane propellers move in a motion that is similar to that of a corkscrew – what is called a helical motion.

An aircraft propeller is an aerodynamic device which converts rotational energy into propulsive force creating thrust which is approximately perpendicular to its plane of rotation. The rotational energy can be produced by a piston or gas turbine engine or, in limited applications, by an electric motor. A propeller can be attached directly to the crankshaft of a piston engine, as is the case in many light aircraft, or it might be powered through a reduction gear box (RGB) attached to a piston or jet engine. In this case, the RGB converts the high rotation speed of the engine to one that is more appropriate for propeller operation. Propellers have two or more blades spaced evenly around the hub and are available in fixed pitch or in variable pitch configurations. More sophisticated propeller designs include those of the constant speed, contra-rotating and counter-rotating types.

Aircraft such as the ATR 72, Tupolev Tu-95, and Bombardier Dash 8 use a system that incorporates propellers powered by engines that are equipped with turbochargers (i.e exhaust gas-driven superchargers), and are hence often referred to as "turboprops." Other aircraft, such as the common single-engine Cessna – 172 use a simple propeller assembly. In this case, the airplane propeller is mounted onto a crankshaft connected to the engine.

## Function of Airplane Propellers

Coordinated balance between the four forces acting on an airplane; lift, weight, thrust and drag, enable an aircraft to fly. Propellers are primarily concerned with the force of thrust – engines deliver their energy output through airplane propellers.

The internal combustion process in the piston engine of aircraft generates power and develops force. In simpler words, the airplane engine provides the propeller with power; the propeller uses this power to create thrust.

Propellers of a C 130J Super Hercules

## Theory and Design of Aircraft Propellers

Lowry quotes a propeller efficiency of about 73.5% at cruise for a Cessna 172. This is derived from his "Bootstrap approach" for analyzing the performance of light general aviation aircraft using fixed pitch or constant speed propellers. The efficiency of the propeller is influenced by the angle of attack ($\alpha$). This is defined as $\alpha = \Phi - \theta$, where $\theta$ is the helix angle (the angle between the resultant relative velocity and the blade rotation direction) and $\Phi$ is the blade pitch angle. Very small pitch and helix angles give a good performance against resistance but provide little thrust, while larger angles have the opposite effect. The best helix angle is when the blade is acting as a wing producing much more lift than drag. However, 'lift-and-drag' is only one way to express the aerodynamic force on the blades. To explain aircraft and engine performance the same force is expressed slightly differently in terms of thrust and torque since the required output of the propeller is thrust. Thrust and torque are the basis of the definition for the efficiency of the propeller as shown below. The advance ratio of a propeller is similar to the angle of attack of a wing.

A propeller's efficiency is determined by

$$\eta = \frac{\text{propulsive power out}}{\text{shaft power in}} = \frac{\text{thrust} \cdot \text{axial speed}}{\text{resistance torque} \cdot \text{rotational speed}}.$$

Propellers are similar in aerofoil section to a low-drag wing and as such are poor in operation when at other than their optimum angle of attack. Therefore, most propellers use a variable pitch mechanism to alter the blades' pitch angle as engine speed and aircraft velocity are changed.

A further consideration is the number and the shape of the blades used. Increasing the aspect ratio of the blades reduces drag but the amount of thrust produced depends on blade area, so using high-aspect blades can result in an excessive propeller diameter. A further balance is that using a smaller number of blades reduces interference effects between the blades, but to have sufficient blade area to transmit the available power within a set diameter means a compromise is needed. Increasing the number of blades also decreases the amount of work each blade is required to perform, limiting the local Mach number – a significant performance limit on propellers. The performance of a propeller suffers when transonic flow first appears on the tips of the blades. As the relative air speed at any section of a propeller is a vector sum of the aircraft speed and the tangential speed due to rotation, the flow over the blade tip will reach transonic speed well before the aircraft does. When the airflow over the tip of the blade reaches its critical speed, drag and torque resistance increase rapidly and shock waves form creating a sharp increase in noise. Aircraft with conventional propellers, therefore, do not usually fly faster than Mach 0.6. There have been propeller aircraft which attained up to the Mach 0.8 range, but the low propeller efficiency at this speed makes such applications rare.

The three-bladed propeller of a light aircraft: the Vans RV-7A

There have been efforts to develop propellers for aircraft at high subsonic speeds. The 'fix' is similar to that of transonic wing design. The maximum relative velocity is kept as low as possible by careful control of pitch to allow the blades to have large helix angles; thin blade sections are used and the blades are swept back in a scimitar shape (Scimitar propeller); a large number of blades are used to reduce work per blade and so circulation strength; contra-rotation is used. The propellers designed are more efficient than turbo-fans and their cruising speed (Mach 0.7–0.85) is suitable for airliners, but the noise generated is tremendous.

## Forces Acting on a Propeller

Forces acting on the blades of an aircraft propeller include the following. Some of these forces can be arranged to counteract each other, reducing the overall mechanical stresses imposed.

## Thrust Bending

Thrust loads on the blades, in reaction to the force pushing the air backwards, act to bend the blades forward. Blades are therefore often *raked* forwards, such that the outward centrifugal force of rotation acts to bend them backwards, thus balancing out the bending effects.

## Centrifugal and Aerodynamic Twisting

A centrifugal twisting force is experienced by any asymmetrical spinning object. In the propeller it acts to twist the blades to a fine pitch. The aerodynamic centre of pressure is therefore usually arranged to be slightly forward of its mechanical centreline, creating a twisting moment towards coarse pitch and counteracting the centrifugal moment. However in a high-speed dive the aerodynamic force can change significantly and the moments can become unbalanced.

## Centrifugal

The force felt by the blades acting to pull them away from the hub when turning. It can be arranged to help counteract the thrust bending force.

## Torque Bending

Air resistance acting against the blades, combined with inertial effects causes propeller blades to bend away from the direction of rotation.

## Vibratory

Many types of disturbance set up vibratory forces in blades. These include aerodynamic excitation as the blades pass close to the wing and fuselage. Piston engines introduce torque impulses which may excite vibratory modes of the blades and cause fatigue failures. Torque impulses are not present when driven by a gas turbine engine.

## Curved Propeller Blades

Since the 1940s, propellers and propfans with swept tips or curved "scimitar-shaped" blades have been studied for use in high-speed applications so as to delay the onset of shockwaves, in similar manner to wing sweepback, where the blade tips approach the speed of sound.

## Varying Pitch

The purpose of varying pitch angle is to maintain an optimal angle of attack for the propeller blades, giving maximum efficiency throughout the flight regime. The requirement for pitch variation is shown by the propeller performance during the Schneider Trophy competition in 1931. The Fairey Aviation Company fixed-pitch propeller used was stalled on take-off up to 160 mph on its way up to a top speed of 407.5 mph. The very wide speed range was achieved because some of the usual requirements for aircraft performance did not apply. There was no compromise on top-speed efficiency, the take-off distance was not restricted to available runway length and there was no climb requirement.

For the highest possible speed the highest possible propeller efficiency is required at the high speed condition. As pitch corresponds to airspeed a coarse pitch is required. The variable pitch blades used on the Tupolev Tu-95propel it at a speed exceeding that once considered possible for a propeller-driven aircraft using an exceptionally coarse pitch.

## Variable Pitch

Early pitch control settings were pilot operated, either with a small number of preset positions or continuously variable.

Following World War I, automatic propellers were developed to maintain an optimum angle of attack. This was done by balancing the centripetal twisting moment on the blades and a set of counterweights against a spring and the aerodynamic forces on the blade. Automatic props had the advantage of being simple, lightweight, and requiring no external control, but a particular propeller's performance was difficult to match with that of the aircraft's power plant.

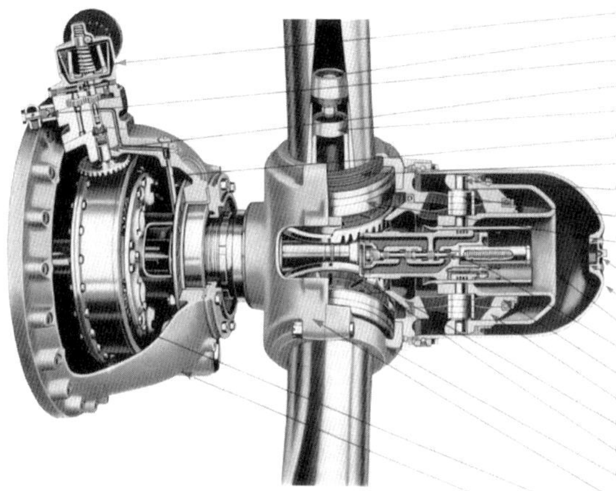


Cut-away view of a Hamilton Standard propeller. This type of constant-speed propeller was used on many American fighters, bombers and transport aircraft of World War II

Modern light aircraft and advanced homebuilt aircraft sometimes have variable pitch (VP) propellers. These tend to be electrically operated and controlled manually or by computer. The V-Prop is self-powering and self-governing.

A simpler version was the spring-loaded "two-speed" VP prop, which was set to fine for takeoff, and then triggered to coarse once in cruise, the propeller then staying in coarse for the remainder of the flight. An even simpler version is the ground-adjustable propeller, which may be adjusted on the ground, but is effectively a fixed-pitch prop once airborne.

## Constant Speed

An improvement on the automatic type was the constant-speed propeller. This type automatically adjusts the blade pitch according to the engine speed, thereby maintaining a constant engine speed for any given manual control setting.Constant-speed propellers allow the pilot to set a rotational speed according to the need for maximum engine power or maximum efficiency, and a propeller governor acts as a closed-loop controller to vary propeller pitch angle as required to maintain the selected engine speed. In most aircraft this system is hydraulic, with engine oil serving as the hydraulic fluid. However, electrically controlled propellers were developed during World War II and saw extensive use on military aircraft, and have recently seen a revival in use on homebuilt aircraft.

## Feathering

On some variable-pitch propellers, the blades can be rotated parallel to the airflow to reduce drag in case of an engine failure. This is called *feathering*, a term borrowed from rowing. On single-engined aircraft, whether a powered glider or turbine-powered aircraft, the effect is to increase the gliding distance. On a multi-engine aircraft, feathering the propeller on a failed engine helps the aircraft maintain altitude with the reduced power from the remaining engines.

Feathered propeller on the outboard TP400 turboprop of an Airbus A400M

Most feathering systems for reciprocating engines sense a drop in oil pressure and move the blades toward the feather position, and require the pilot to pull the propeller control back to disengage the high-pitch stop pins before the engine reaches idle RPM. Turboprop control systems usually utilize a *negative torque sensor* in the reduction gearbox which moves the blades toward feather when the engine is no longer providing power to the propeller. Depending on design, the pilot may have to push a button to override the high-pitch stops and complete the feathering process, or the feathering process may be totally automatic.

## Reverse Pitch

The propellers on some aircraft can operate with a negative blade pitch angle, and thus reverse the thrust from the propeller. This is used to help slow the plane down after landing in order to save wear on the brakes and tires, but in some cases also allows the aircraft to back up on its own – this is particularly useful for getting floatplanes out of confined docks. This is known as Beta Pitch.

## Counter-rotating Propellers

Counter-rotating propellers are sometimes used on twin-engine and multi-engine aircraft with wing-mounted engines. These propellers turn in opposite directions from their counterpart on the other wing to balance out the torque and p-factor effects. They are sometimes referred to as "handed" propellers since there are left hand and right hand versions of each prop.

Generally, the propellers on both engines of most conventional twin-engined aircraft spin clockwise (as viewed from the rear of the aircraft). To eliminate the critical engine problem, counter-rotating propellers usually spin "inwards" towards the fuselage – clockwise on the left engine and counter-clockwise on the right – but there are exceptions such as the P-38 Lightning which spun "outwards" away from the fuselage, and the Airbus A400 whose inboard and outboard engines turn in opposite directions even on the same wing.

## Contra-rotating Propeller

A contra-rotating propeller or contra-prop places two counter-rotating propellers on concentric drive shafts so that one sits immediately 'downstream' of the other propeller. This provides the benefits of counter-rotating propellers for a single powerplant. The forward propeller provides the majority of the thrust, while the rear propeller also recovers energy lost in the swirling motion of the air in the propeller slipstream. Contra-rotation also increases the ability of a propeller to absorb power from a given engine, without increasing propeller diameter. However the added cost, complexity, weight and noise of the system rarely make it worthwhile and it is only used on high-performance types where ultimate performance is more important than efficiency.

Contra-rotating propellers of a modified North American P-51 Mustang fitted with a Rolls-Royce Griffon engine.

## Aircraft Fans

A fan is a propeller with a large number of blades. A fan therefore produces a lot of thrust for a given diameter but the closeness of the blades means that each strongly affects the flow around the others. If the flow is supersonic, this interference can be beneficial if the flow can be compressed through a series of shock waves rather than one. By placing the fan within a shaped duct, specific flow patterns can be created depending on flight speed and engine performance. As air enters the duct, its speed is reduced while its pressure and temperature increase. If the aircraft is at a high subsonic speed this creates two advantages: the air enters the fan at a lower Mach speed; and the higher temperature increases the local speed of sound. While there is a loss in efficiency as the fan is drawing on a smaller area of the free stream and so using less air, this is balanced by the ducted fan retaining efficiency at higher speeds where conventional propeller efficiency would be poor. A ducted fan or propeller also has certain benefits at lower speeds but the duct needs to be shaped in a different manner than one for higher speed flight. More air is taken in and the fan therefore operates at an efficiency equivalent to a larger un-ducted propeller. Noise is also reduced by the ducting and should a blade become detached the duct would help contain the damage. However the duct adds weight, cost, complexity and (to a certain degree) drag.

## Empennage

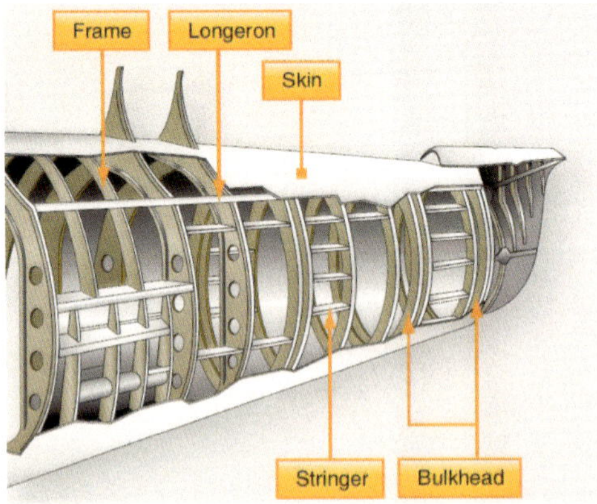

The fuselage terminates at the tail cone with similar but more lightweight construction

The empennage of an aircraft is also known as the tail section. Most empennage designs consist of a tail cone, fixed aerodynamic surfaces or stabilizers, and movable aerodynamic surfaces. The tail cone serves to close and streamline the aft end of most fuselages. The cone is made up of structural members like those of the fuselage; however, cones are usually of lighter construction since they receive less stress than the fuselage.

The other components of the typical empennage are of heavier construction than the tail cone. These members include fixed surfaces that help stabilize the aircraft and movable surfaces that help to direct an aircraft during flight. The fixed surfaces are the horizontal stabilizer and vertical stabilizer. The movable surfaces are usually a rudder located at the aft edge of the vertical stabilizer and an elevator located at the aft edge the horizontal stabilizer.

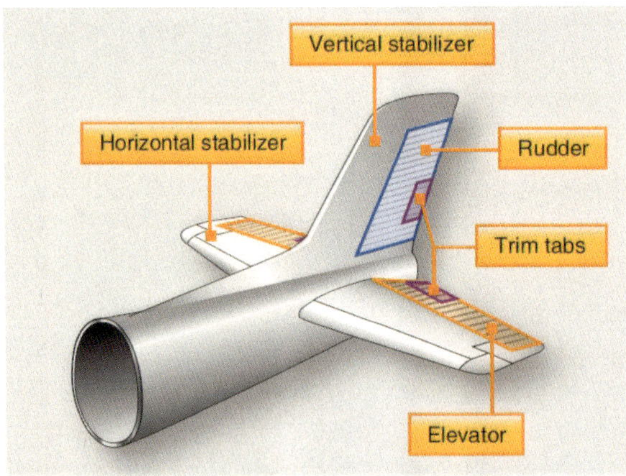

Components of a typical empennage

The structure of the stabilizers is very similar to that which is used in wing construction. shows a typical vertical stabilizer. Notice the use of spars, ribs, stringers, and skin like those found in a wing. They perform the same functions shaping and supporting the stabilizer and transferring stresses. Bending, torsion, and shear created by air loads in flight pass from one structural member to another. Each member absorbs some of the stress and passes the remainder on to the others. Ultimately, the spar transmits any overloads to the fuselage. A horizontal stabilizer is built the same way.

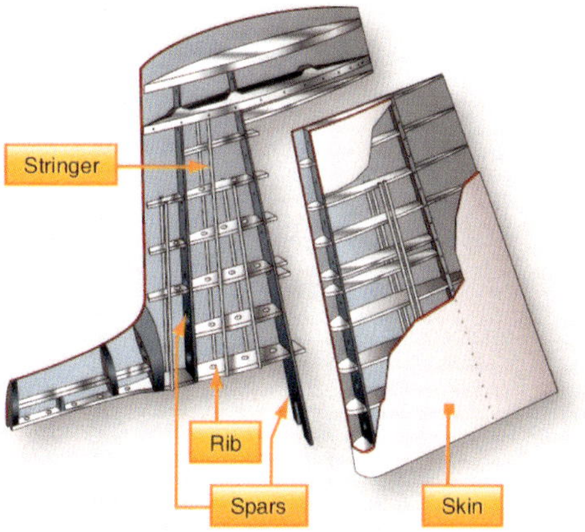


Vertical stabilizer

## Structure

Structurally, the empennage consists of the entire tail assembly, including the tailfin, the tailplane and the part of the fuselage to which these are attached. On an airliner this would be all the flying and control surfaces behind the rear pressure bulkhead.

The front (usually fixed) section of the tailplane is called the tailplane or horizontal stabiliser and is used to provide pitch stability. The rear section is called the elevator, and is usually hinged to the horizontal stabiliser. The elevator is a movable aerofoil that controls changes in pitch, the up-and-down motion of the aircraft's nose. Some aircraft employ an all-moving stabiliser and elevators in one unit, known as a stabilator or "full-flying stabiliser".

The vertical tail structure (or fin) has a fixed front section called the vertical stabiliser, used to restrict side-to-side motion of the aircraft (yawing). The rear section of the vertical fin is the rudder, a movable aerofoil that is used to turn the aircraft's nose to one side or the other. When used in combination with the ailerons, the result is a banking turn, often referred to as a "coordinated turn".

Some aircraft are fitted with a tail assembly that is hinged to pivot in two axes forward of the fin and stabiliser, in an arrangement referred to as a *movable tail*. The entire empennage is rotated vertically to actuate the horizontal stabiliser, and sideways to actuate the fin.

The aircraft's cockpit voice recorder, flight data recorder and Emergency locator transmitter (ELT) are often located in the empennage, because the aft of the aircraft provides better protection for these in most aircraft crashes.

## Trim

In some aircraft trim devices are provided to eliminate the need for the pilot to maintain constant pressure on the elevator or rudder controls.

The trim device may be:

- A trim tab on the rear of the elevators or rudder which act to change the aerodynamic load on the surface. Usually controlled by a cockpit wheel or crank.

- An adjustable stabiliser into which the stabiliser may be hinged at its spar and adjustably jacked a few degrees in incidence either up or down. Usually controlled by a cockpit crank.
- A bungee trim system which uses a spring to provide an adjustable preload in the controls. Usually controlled by a cockpit lever.
- An anti-servo tab used to trim some elevators and stabilators as well as increased control force feel. Usually controlled by a cockpit wheel or crank.
- A servo tab used to move the main control surface, as well as act as a trim tab. Usually controlled by a cockpit wheel or crank.

Multi-engined aircraft often have trim tabs on the rudder to reduce the pilot effort required to keep the aircraft straight in situations of asymmetrical thrust, such as single engine operations.

## Tail Configurations

Aircraft empennage designs may be classified broadly according to the fin and tailplane configurations.

The overall shapes of individual tail surfaces (tailplane planforms, fin profiles) are similar to wing planforms.

## Tailplanes

The tailplane comprises the tail-mounted fixed horizontal stabiliser and movable elevator. Besides its planform, it is characterised by:

- Number of tailplanes - from 0 (tailless or canard) to 3 (Roe triplane).
- Location of tailplane - mounted high, mid or low on the fuselage, fin or tail booms.
- Fixed stabiliser and movable elevator surfaces, or a single combined stabilator or *flying tail.* (General Dynamics F-111 Aardvark).

Some locations have been given special names:

- Cruciform tail - The horizontal stabilisers are placed midway up the vertical stabiliser, giving the appearance of a crosswhen viewed from the front. Cruciform tails are often used to keep the horizontal stabilisers out of the engine wake, while avoiding many of the disadvantages of a T-tail. Examples include the Hawker Sea Hawk and Douglas A-4 Skyhawk.
- T-tail - The horizontal stabiliser is mounted on top of the fin, creating a "T" shape when viewed from the front. T-tails keep the stabilisers out of the engine wake, and give better pitch control. T-tails have a good glide ratio, and are more efficient on low speed aircraft. However, T-tails are more likely to enter a deep stall, and are more difficult to recover from a spin. T-tails must be stronger, and therefore heavier than conventional tails. T-tails also have a larger radar cross section. Examples include the Gloster Javelin, Boeing 727 and McDonnell Douglas DC-9.

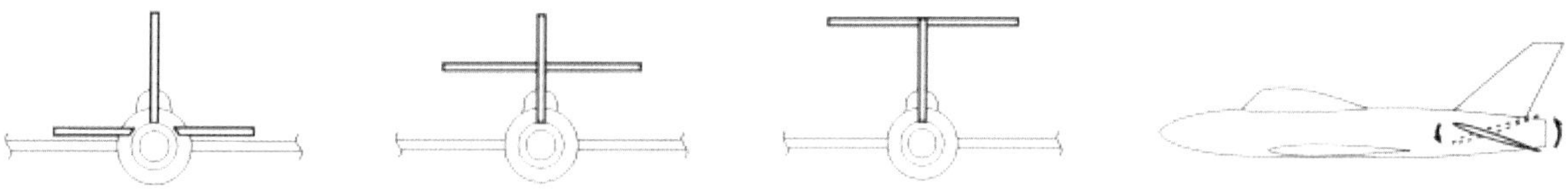
Fuselage mounted | Cruciform | T-tail | Flying tailplane

## Fins

The fin comprises the fixed vertical stabiliser and rudder. Besides its profile, it is characterised by:

- Number of fins - usually one or two.
- Location of fins - on the fuselage (over or under), tailplane, tail booms or wings.

Twin fins may be mounted at various points:

- Twin tail A twin tail, also called an H-tail, consists of two small vertical stabilisers on either side of the horizontal stabiliser. Examples include the Antonov An-225 Mriya, B-25 Mitchell, Avro Lancaster, and ERCO Ercoupe.
- Twin boom A twin boom has two fuselages or booms, with a vertical stabiliser on each, and a horizontal stabiliser between them. Examples include the P-38 Lightning, de Havilland Vampire, Sadler Vampire, and Edgley Optica.
- Wing mounted midwing as on the F7U Cutlass or on the wing tips as on the Handley Page Manx and Rutan Long-EZ.

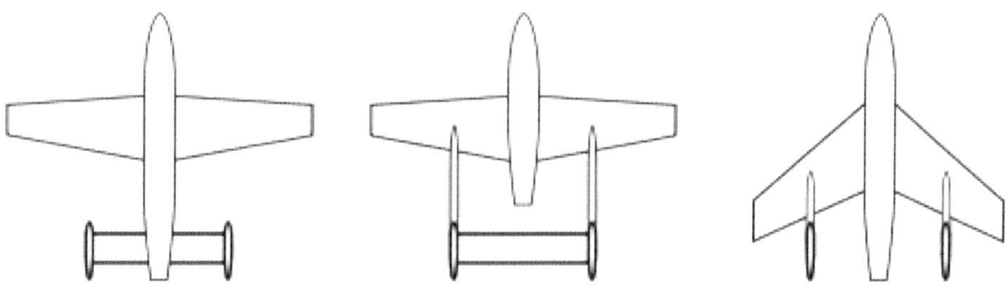

| Tailplane mounted | Twin tail boom | Wing mounted |
|---|---|---|

Unusual fin configurations include:

- No fin - as on the McDonnell Douglas X-36. This configuration is sometimes incorrectly referred to as “tailless”.
- Multiple fins - examples include the Lockheed Constellation (three), Bellanca 14-13 (three), and the Northrop Grumman E-2 Hawkeye (four).
- Ventral fin - underneath the fuselage. Often used in addition to a conventional fin as on the (North American X-15 and Dornier Do 335).

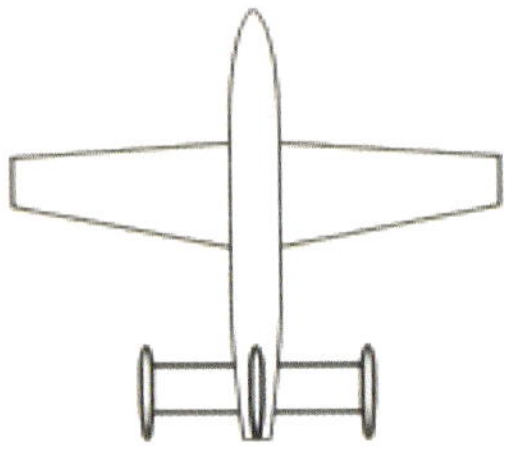

Triple fins

Ventral fin

## V and X Tails

An alternative to the fin-and-tailplane approach is provided by the V-tail and X-tail designs. Here, the tail surfaces are set at diagonal angles, with each surface contributing to both pitch and yaw. The control surfaces, sometimes called ruddervators, act differentially to provide yaw control (in place of the rudder) and act together to provide pitch control (in place of the elevator).

- V tail: A V-tail can be lighter than a conventional tail in some situations and produce less drag, as on the Fouga Magistertrainer, Northrop Grumman RQ-4 Global Hawk RPV and X-37 spacecraft. A V-tail may also have a smaller radar signature. Other aircraft featuring a V-tail include the Beechcraft Model 35 Bonanza, and Davis DA-2. A slight modification to the V-tail can be found on the Waiex and Monnett Moni called a Y-tail.
- Inverted V tail: The unmanned Predator uses an inverted V-tail as do the Lazair and Mini-IMP.
- X tail: The Lockheed XFV and Convair XFY Pogo both featured "X" tails, which were reinforced and fitted with a wheel on each surface so that the craft could sit on its tail and take off and land vertically.

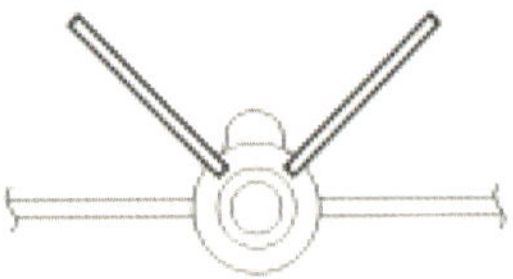

V-tail

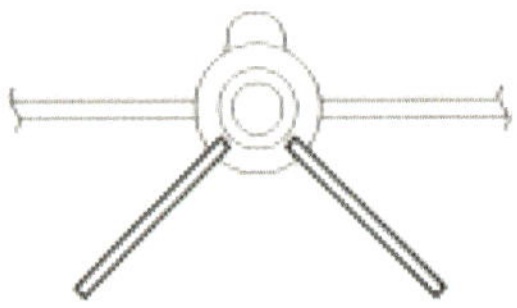

Inverted V-tail

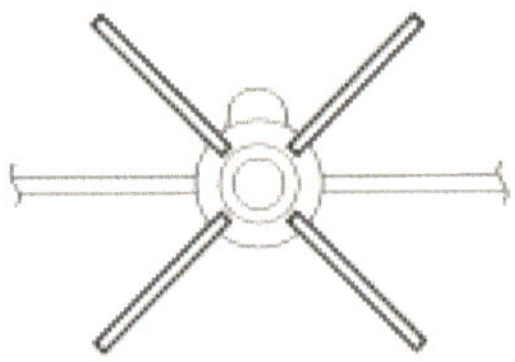

X-tail

## Outboard Tail

SpaceShipOne at the US National Air and Space Museum

An outboard tail is split in two, with each half mounted on a short boom just behind and outboard of each wing tip. It comprises outboard horizontal stabilizers (OHS) and may or may not include additional boom-mounted vertical stabilizers (fins). In this position, the tail surfaces interact constructively with the wingtip vortices and, with careful design, can significantly reduce drag to improve efficiency, without adding unduly to the structural loads on the wing.

The configuration was first developed during World War II by Richard Vogt and George Haag at Blohm & Voss. The Skoda-Kauba SL6 tested the proposed control system in 1944 and, following several design proposals, an order was received for the Blohm & Voss P 215 just weeks before the war ended. The outboard tail reappeared on the Scaled Composites SpaceShipOne in 2003 and SpaceShipTwo in 2010.

### Tailless Aircraft

The DH108 *Swallow*

A tailless aircraft (often tail-less) traditionally has all its horizontal control surfaces on its main wing surface. It has no horizontal stabiliser - either tailplane or canardforeplane (nor does it have a second wing in tandem arrangement). A 'tailless' type usually still has a vertical stabilising fin (vertical stabiliser) and control surface (rudder). However, NASA adopted the 'tailless' description for the novel X-36 research aircraftwhich has a canard foreplane but no vertical fin.

The most successful tailless configuration has been the tailless delta, especially for combat aircraft.

## Aircraft Design Process

There are a few key principles which can help guide engineers through any design process, through improved analysis the fundamentals of design, this same process can be refined into a aircraft design process. However, before moving on to design process, it is important that an engineering product follows a very similar design approach.

### The General Engineering Design Process

There are two types of engineering companies; assembler companies and the sub-contractors. Companies like Jaguar Land Rover, Airbus, and Apple are the assemblers. They are the brand you buy. But these companies often outsource their work to sub-contractors: smaller companies developing specialized products.

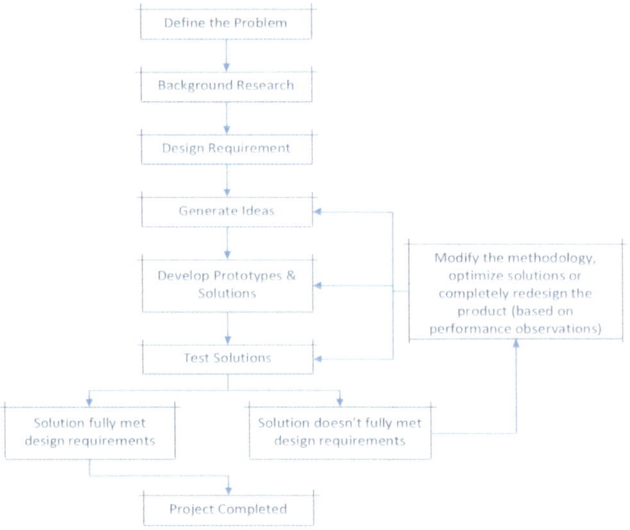


Standard Engineering Process flow Map

A very typical approach to any product development.

## The Aircraft Design Process

The design process for an aircraft can be seen as very similar to other engineering products. However, the complexity of an aircraft and the tight safety regulations required make it a very expensive and long process. The general process can be divided as shown below:

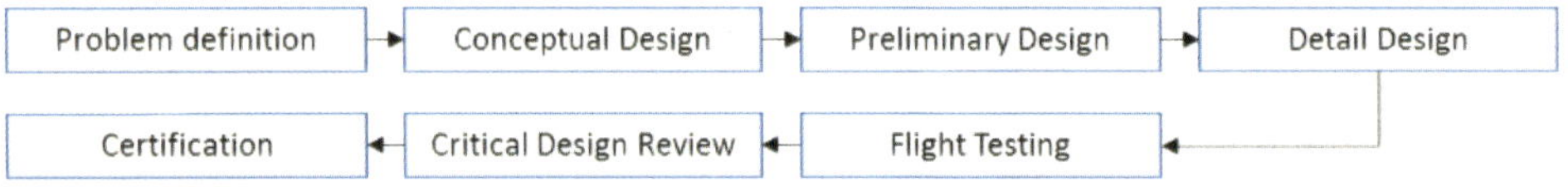


Aircraft design: process-map Example

In order to succeed, it is important to properly frame your issue by identifying key design requirements. Creating a process map is the first step towards designing and building a new aircraft.

Step 1: Problem Definition

At this initial stage, a new market demand has been identified. For example, a company might have identified that there is a growing demand for small private jets. In the inital problem definition stages it is crutial to first identify the key requirements of the new product then to determine how realistic it is to create this product through a feasibility analysis.

Step 2: Conceptual Design

This stage includes back of the envelope calculation, where all the ideas are brainstorm. As the design continues to be revised it will undergoe many transformations to conform to the fit/form/functions that are desired by the end user or purchaser. These basic functions are brought together to create the initial configuration design which is used to discuss different possibilities with the consumer during a conceptual design review.

Step 3: Preliminary Design

The preliminary design phase is where engineers go a step further into the design process to create a basic proof of concept. In this stage the engineers will calculate requirements that the airplane will have to conform with in order to fly, these can include Aerodynamics, Flight Mechanics, Structure stresses and stability. Following these is a preliminary design review, to ensure the concept can become a reality and for a reasonable price.

Step 4: Detail Design

At the detail design stage, every single component will be designed from and tested. This is a highly complex and critical stage of the process. It also requires all the different design teams to work closely. The main aircraft components designed stages are:

1. Wing Design
2. Tail Design
3. Fuselage Design
4. Propulsion System
5. Landing Gear
6. Equipment/Subsystems
7. Integration
8. Wing Tunnel Testing
9. Wight Distribution
10. Performance and stability analysis
11. Control Surfaces design
12. Evaluation and test review

Step 5: Flight Testing

At this stage, a prototype is build and will undergo rigorous flight tests.This is a very lengthy and expensive process to ensure quality and safety in the new aircraft, this can take years to complete. There are two different types of aircraft certifications:

Military Aircraft: e.g.: Mil-Specs

Civil Aircraft: e.g.: EASA (Europe) or FAA (USA)

In order to become certified the design must follow these standards, sometimes also referred to as Technical Standard Orders (TSOs) which govern the approval of the design.

Step 6: Critical Design Review

Here the design is reviewed, final small changes are made to adjust for different needs identified in flight testing.

Step 7: Certification

Once the aircraft is certified, it can now be allowed into commercial airspace and move into full use and production. During the final certification process there are a wide variety of approvals that are needed to be obtained for design and safety:

1. Aircraft Software
2. Engines, Propellers, Aux power units
3. Human safety factors
4. Technical Standard Orders
5. Parts Manufacturer Approval
6. Origional Design Approval Process
7. National Automated Conformity Inspection Process

This myriad of tests and approvals is done to ensure the safe operation and functions of the aircrafts.

## Different Versions of Aircrafts

Okay, so you have noticed that there are several versions of the Airbus A320 or even Boeing 777. Well, designing an aircraft from scratch is a very long process (15-20 years) and extremely expensive. Hence a solution to this is to modify existing aircraft by upgrading some parts. These newer versions are then released under a different variant name. These can be done much more quickly as most of the aircraft has undergone the approval and certification processes already, this is why you have the Airbus A320 family.

## References


- Mesko, Jim (2002). FH Phantom/F2H Banshee in action. Carrollton, Texas, USA: Squadron/Signal Publications, Inc. p. 12. ISBN 0-89747-444-9
- Florence Graves and Sara K. Goo (Apr 17, 2006). "Boeing Parts and Rules Bent, Whistle-Blowers Say". Washington Post. Retrieved April 23, 2010

- Brooks, Peter, W., Zeppelin: Rigid Airships 1893–1940, Washington, Smithsonian Institution Press, 1992, ISBN 1-56098-228-4 p. 19
- S. Ulamec; S. Espinasse; B. Feuerbacher; M. Hilchenbach; D. Moura; H. Rosenbauer; H. Scheuerle; R. Willnecker (2006). "Rosetta Lander—Philae: Implications of an alternative mission". Acta Astronautica. 58 (8): 435–441. Bibcode:2006AcAau..58..435U. doi:10.1016/j.actaastro.2005.12.009
- Soffen, G.A., Snyder, C.W. (August 1976). "The First Viking Mission to Mars". Science, New Series. 193 (4255): 759–766. Bibcode:1976Sci...193..759S. doi:10.1126/science.193.4255.759. JSTOR 1742875
- Sara Kehaulani Goo (Oct 27, 2004). "NTSB Cites Pilot Error in 2001 N.Y. Crash". Washington Post. Retrieved April 23, 2010
- E. V. Laitone, Wind tunnel tests of wings at Reynolds numbers below 70 000, Experiments in Fluids 23, 405 (1997). doi:10.1007/s003480050128
- Kundu, Ajoy Kumar (2010). Aircraft Design. New York: Cambridge University Press. p. 194. ISBN 9780511677854
- Stern, Jonathan M. (1995). "Chapter 2: The New Flight Simulator Pilots Ground School: Retractable Landing Gear". Microsoft Flight Simulator Handbook. Indianapolis, Indiana: Brady Publishing. ISBN 1566862825. Retrieved 13 May 2012
- Sengfelder, Günther (1993). German Aircraft Landing Gear. Atglen, PA USA: Schiffer Publishing. pp. 40–42. ISBN 0-88740-470-7
- Dow, Andrew (February 19, 2015). Pegasus: The Heart of the Harrier (2 ed.). Pen and Sword. p. 312. ISBN 1473827604. Retrieved September 6, 2017
- J. Biele (2002). "The Experiments Onboard the ROSETTA Lander". Journal Earth, Moon, and Planets. 90 (1–4): 445–458. Bibcode:2002EM&P...90..445B. doi:10.1023/A:1021523227314
- Air Transport Association (10 November 2011). "ATA Airline Handbook Chapter 5: How Aircraft Fly". Archived from the original on 10 November 2011. Retrieved 5 March 2013

# Permissions

All chapters in this book are published with permission under the Creative Commons Attribution Share Alike License or equivalent. Every chapter published in this book has been scrutinized by our experts. Their significance has been extensively debated. The topics covered herein carry significant information for a comprehensive understanding. They may even be implemented as practical applications or may be referred to as a beginning point for further studies.

We would like to thank the editorial team for lending their expertise to make the book truly unique. They have played a crucial role in the development of this book. Without their invaluable contributions this book wouldn't have been possible. They have made vital efforts to compile up to date information on the varied aspects of this subject to make this book a valuable addition to the collection of many professionals and students.

This book was conceptualized with the vision of imparting up-to-date and integrated information in this field. To ensure the same, a matchless editorial board was set up. Every individual on the board went through rigorous rounds of assessment to prove their worth. After which they invested a large part of their time researching and compiling the most relevant data for our readers.

The editorial board has been involved in producing this book since its inception. They have spent rigorous hours researching and exploring the diverse topics which have resulted in the successful publishing of this book. They have passed on their knowledge of decades through this book. To expedite this challenging task, the publisher supported the team at every step. A small team of assistant editors was also appointed to further simplify the editing procedure and attain best results for the readers.

Apart from the editorial board, the designing team has also invested a significant amount of their time in understanding the subject and creating the most relevant covers. They scrutinized every image to scout for the most suitable representation of the subject and create an appropriate cover for the book.

The publishing team has been an ardent support to the editorial, designing and production team. Their endless efforts to recruit the best for this project, has resulted in the accomplishment of this book. They are a veteran in the field of academics and their pool of knowledge is as vast as their experience in printing. Their expertise and guidance has proved useful at every step. Their uncompromising quality standards have made this book an exceptional effort. Their encouragement from time to time has been an inspiration for everyone.

The publisher and the editorial board hope that this book will prove to be a valuable piece of knowledge for students, practitioners and scholars across the globe.

# Index